Module C3 — Chemicals in Our Lives

Tectonic Plates 56
Resources in the Earth's Crust 57
Salt 58
Salt in the Food Industry 59
Electrolysis of Salt Solution 60
Chlorination 61
Alkalis 62
Impacts of Chemical Production 63
Life Cycle Assessments 64

Module P1 — The Earth in the Universe

The Solar System 65
Beyond the Solar System 66
Looking Into Space 67
The Life of the Universe 68
The Changing Earth 69
Wegener's Theory of Continental Drift 70
The Structure of the Earth 71
Seismic Waves 72
Waves — The Basics 73

Module P2 — Radiation and Life

Electromagnetic Radiation 75
Electromagnetic Radiation and Energy 76
Ionising Radiation 77
Microwaves 78
Electromagnetic Radiation and the Atmosphere 79
The Carbon Cycle 80
Global Warming and Climate Change 81
Electromagnetic Waves and Communication 82
Analogue and Digital Signals 83

Module P3 — Sustainable Energy

Electrical Energy 84
Efficiency 86
Sankey Diagrams 87
Saving Energy 88
Energy Sources and Power Stations 89
Nuclear Energy 90
Wind and Solar Energy 91
Wave and Tidal Energy 92
Biofuels and Geothermal Energy 93
Hydroelectricity and Reliable Fuel Supplies 94
Comparing Energy Resources 95
Generators and the National Grid 96

Controlled Assessment — Practical Data Analysis

Planning 97
Processing the Data 98
Conclusion and Evaluation 100
The Case Study 101

Answers 102
Index 107

GREENWICH LIBRARIES	
ELT	
3 8028 02249841 1	
Askews & Holts	11-May-2016
C507.6	£5.95
5028666	

Published by CGP

Editors:
Ellen Bowness, Katie Braid, Emma Elder, Ben Fletcher, Murray Hamilton, Edmund Robinson,
Helen Ronan, Jane Towle, Julie Wakeling, Karen Wells, Dawn Wright.

ISBN: 978 1 84762 719 3

With thanks to Katherine Craig and Hayley Thompson for the proofreading.

With thanks to Janet Cruse-Sawyer, Ian Francis, Lorna Henderson
and Jamie Sinclair for the reviewing.

With thanks to Jan Greenway, Laura Jakubowski and Laura Stoney for the copyright research.

With thanks to Science Photo Library for permission to use the image on page 47.

With thanks to iStockphoto.com for permission to use the image on page 69.

Groovy website: www.cgpbooks.co.uk

Printed by Elanders Ltd, Newcastle upon Tyne.
Jolly bits of clipart from CorelDRAW®

Based on the classic CGP style created by Richard Parsons.

Photocopying — it's dull, grey and sometimes a bit naughty. Luckily, it's dead cheap, easy and quick to order
more copies of this book from CGP — just call us on 0870 750 1242. Phew!
Text, design, layout and original illustrations © Coordination Group Publications Ltd. (CGP) 2011
All rights reserved.

How to get your free Online Edition

Want to read this book on a computer or tablet? Just go to **cgpbooks.co.uk/extras** and enter this code...

0950 4945 6248 8884

By the way, this code only works for one person. If somebody else has used this book before you, they might have already claimed the online extras.

GCSE OCR 21st Century
Core Science
Foundation — the Basics
The Revision Guide

This book is for anyone doing **GCSE OCR 21st Century Core Science** at Foundation Level, with a predicted grade of D or below.
(If you're not sure what your predicted grade is, your teacher will be able to tell you.)

All the important topics are explained in a clear, straightforward way to help you get all the marks you can in the exam.

And of course, there are some daft bits to make the whole thing vaguely entertaining for you.

What CGP is all about

Our sole aim here at CGP is to produce the highest quality books — carefully written, immaculately presented and dangerously close to being funny.

Then we work our socks off to get them out to you — at the cheapest possible prices.

Contents

	Course Structure 1

Ideas About Science

- The Scientific Process 2
- Data 4
- Correlation and Cause 5
- Risk 6
- Science and Ethics 7

Using Equations 8

Module B1 — You and Your Genes

- Genes, Chromosomes and DNA 9
- Genes and Variation 10
- Inheritance and Genetic Diagrams 11
- Genetic Diagrams and Sex Chromosomes 13
- Genetic Disorders 14
- Genetic Testing 15
- Clones 16
- Stem Cells 17

Module B2 — Keeping Healthy

- Microorganisms and Disease 18
- The Immune System 19
- Vaccination 20
- Antibiotics 21
- Drug Trials 22
- The Circulatory System 23
- Heart Rate and Blood Pressure 24
- Heart Disease 25
- Homeostasis and The Kidneys 26
- Controlling Water Content 27

Module B3 — Life on Earth

- Adaptation and Variation 28
- Natural Selection and Selective Breeding 29
- Evolution 30
- Biodiversity and Classification 32
- Energy in an Ecosystem 33
- Interactions Between Organisms 35
- The Carbon Cycle 36
- The Nitrogen Cycle 37
- Measuring Environmental Change 38
- Sustainability 39

Module C1 — Air Quality

- How the Air was Made 41
- The Air Today 42
- Chemical Reactions 43
- Fuels 44
- Air Pollution — Carbon 45
- Air Pollution — Sulfur and Nitrogen 46
- Reducing Pollution 47

Module C2 — Material Choices

- Natural and Synthetic Materials 48
- Materials and Properties 49
- Materials, Properties and Uses 50
- Crude Oil 51
- Uses of Crude Oil 52
- Polymerisation 53
- Structure and Properties of Polymers 54
- Nanotechnology 55

Course Structure

You need to know <u>what exams</u> you're taking and <u>what to learn</u> for each one.
It may seem dull as dishwater, but it's <u>really important</u>. So read on...

There are Different Parts to GCSE Science

1) In your exams you'll be tested on <u>Biology</u>, <u>Chemistry</u> and <u>Physics</u>. These are all covered in this book.
2) You need to know about <u>Ideas About Science</u> too. There's a <u>whole section</u> on this — see pages 2-7.
3) You also have to do a <u>Controlled Assessment</u> — it's a bit like a <u>coursework exam</u>.
   There's a <u>whole section</u> on pages 97-101 to <u>help you</u> with the Controlled Assessment.

There are Two Different Sets of Exams You Could Do

There are <u>two different ways</u> you can be tested for GCSE Science — <u>Route 1</u> and <u>Route 2</u>.
Your <u>teacher</u> will be able to tell you which one you're doing.

If You're Doing Route 1...

You'll have to do <u>three</u> exams...

You need to revise Ideas About Science for all of these exams too.

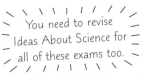

1) The <u>BIOLOGY</u> exam.
   You need to revise <u>Modules B1, B2 and B3</u> of this book for this exam.

2) The <u>CHEMISTRY</u> exam.
   You need to revise <u>Modules C1, C2 and C3</u> of this book for this exam.

3) The <u>PHYSICS</u> exam.
   You need to revise <u>Modules P1, P2 and P3</u> of this book for this exam.

You also have to do a <u>Controlled Assessment</u>.

If You're Doing Route 2...

You'll have to do <u>three</u> exams...

You need to revise Ideas About Science for all of these exams too.

1) The <u>Unit 1 exam</u> tests you on <u>BIOLOGY</u>, <u>CHEMISTRY</u> and <u>PHYSICS</u>.
   You need to revise <u>Modules B1, C1 and P1</u> of this book for this exam.

2) The <u>Unit 2 exam</u> tests you on <u>BIOLOGY</u>, <u>CHEMISTRY</u> and <u>PHYSICS</u> as well.
   You need to revise <u>Modules B2, C2 and P2</u> of this book for this exam.

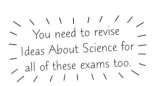

3) The <u>Unit 3 exam</u> tests you on <u>BIOLOGY</u>, <u>CHEMISTRY</u> and <u>PHYSICS</u> too.
   You need to revise <u>Modules B3, C3 and P3</u> of this book for this exam.

You also have to do a <u>Controlled Assessment</u>.

Ideas About Science

The Scientific Process

Scientists try to find <u>explanations</u> for things that happen. Here's the way that <u>most scientists</u> work...

A <u>Hypothesis</u> is an <u>Explanation of Something</u>

1) Scientists <u>OBSERVE</u> (look at) things they <u>don't understand</u>.
2) They then come up with an <u>explanation</u> for what they've seen.
3) This explanation is called a <u>HYPOTHESIS</u>.

For Example:

A scientist is looking at <u>why</u> people have <u>spots</u>.

He notices that everyone with spots <u>picks their nose</u>.

The scientist thinks that the spots might be <u>caused</u> by people picking their nose.

So the <u>hypothesis</u> is: "People have spots because they pick their nose."

Different Scientists <u>Can Come Up With</u> Different Explanations

1) Different scientists can observe the <u>same things</u> and come up with <u>DIFFERENT EXPLANATIONS</u> (different <u>hypotheses</u>) for them.
2) This is because <u>different scientists</u> might <u>think in different ways</u>.
3) This means it's important to <u>test</u> the explanations.
4) Then you can see which one is <u>most likely to be TRUE</u>.

Scientists <u>Test the Hypothesis</u>

1) Scientists <u>test</u> whether a <u>hypothesis</u> is <u>RIGHT or NOT</u> by making a <u>PREDICTION</u> and <u>TESTING</u> it.

For Example:

A prediction is something like: "People who pick their nose will have spots."

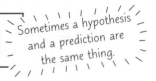

Sometimes a hypothesis and a prediction are the same thing.

2) If tests show that the <u>prediction</u> is <u>RIGHT</u>, then there's <u>EVIDENCE</u> (signs) that the <u>hypothesis is right</u> too.
3) If tests show that the <u>prediction</u> is <u>WRONG</u>, then the <u>hypothesis</u> is probably <u>wrong</u> as well.

Ideas About Science

The Scientific Process

The scientific process can be quite long... which is why there's another page on it.
I bet you just can't wait to see how it ends up. Enjoy.

Other Scientists Test the Hypothesis by Making Predictions

1) It's NOT enough for one scientist to do tests to see if the hypothesis is right or not.
2) OTHER scientists test the hypothesis as well.
3) They try to get the same results as the first scientist.
4) They also try to find MORE EVIDENCE that the hypothesis is RIGHT.
5) If they DO get the same results or find more evidence then the hypothesis is ACCEPTED. It then goes into books for people to learn.

When scientists check each other's work, it's called PEER REVIEW.

6) Sometimes the scientists will find evidence that shows the hypothesis is WRONG.
7) When this happens the scientists have to start all over again. Sad times.

If Evidence Supports a Hypothesis, It's Accepted — for Now

1) If a hypothesis gets through peer review, then scientists start to trust it a lot. They then call it a theory.

2) Once scientists have accepted a theory, they take a lot of persuading to drop it.

3) The tried and tested theory will stay until a better explanation is found.

Ideas About Science

Data

This page is all about what scientists do with <u>data</u> (their results) and why it's so <u>important</u>...

Scientists Need Reliable Data

1) The only way to <u>test a hypothesis</u> is to collect <u>RELIABLE data</u>.
2) If a scientist does an investigation <u>lots of times</u> and gets the <u>SAME DATA each time</u> then it's <u>reliable</u>.
3) It's also reliable if <u>OTHER SCIENTISTS</u> get the <u>same data</u> when they do the investigation.

Data Will Always Vary

1) If you take a lot of measurements of the <u>same thing</u>, you <u>WON'T</u> always get the <u>SAME RESULT</u>.
2) This might be because you made a <u>mistake</u> when measuring.
3) This means you <u>can't be sure</u> that just <u>one measurement</u> will give you the <u>REAL VALUE</u>.

Repeating Measurements Helps You Get Reliable Results

Scientists <u>repeat</u> their investigations to make sure their results are <u>RELIABLE</u>.

For Example: You want to know how long a reaction takes at 10 °C, 20 °C and 30 °C

Do the reaction <u>THREE TIMES</u> at <u>EACH</u> temperature.
Time <u>how long</u> it takes <u>each time</u> you do the reaction.

Reaction time at <u>10 °C</u>

Reaction time at <u>20 °C</u>

Reaction time at <u>30 °C</u>

All the results are <u>similar</u> so the data is <u>reliable</u>.

Some Results Can be Outliers

1) The results of investigations always <u>vary a bit</u>.
2) But sometimes you get a result that <u>doesn't seem to FIT IN</u> with the rest at all.
3) These results are called <u>OUTLIERS</u>.
4) You should try to find out what <u>caused them</u>.
5) If you can find out, you can <u>IGNORE</u> them when you're looking at the results.

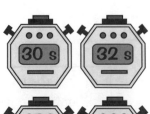
outlier

Ideas About Science

Correlation and Cause

Correlation and cause come up a lot in science. It's important that you understand the difference.

A Correlation is a Relationship Between Two Things

1) Scientists collect data and use it to look for relationships (links) between things.
2) For example, one thing might increase while another thing decreases. Or two things might increase or decrease together.
3) This link is called a CORRELATION.
4) BUT a correlation DOESN'T mean that one thing CAUSES the other.

There's more on correlation on page 99.

- For example, a scientist finds that the temperature of the sea and the number of shark attacks both increase together.
- BUT this DOESN'T mean that sharks are more likely to attack in warm water — it's just that there are more people in the water to attack when it's warm.

Investigations Have to be Fair Tests

1) Scientists need to make sure that their investigations are FAIR TESTS.
2) To make their investigation a fair test, the scientist must...

ONLY CHANGE ONE THING. EVERYTHING ELSE must be kept the SAME.

3) This means that the scientist can tell whether one thing is causing another.

Example: Investigation to see how changing the amount of light changes how tall a plant grows

Change the amount of light the plant gets...

"Anybody out there...?"

...but keep everything else the same.

Same temperature   Same type of plant   Same amount of water

Ideas About Science

Risk

Reading this page has <u>risks</u>. Paper cuts, falling asleep... Science has lots of risks too.

Nothing is Risk-Free

1) <u>Everything</u> that you do has a <u>RISK</u> attached to it.
2) <u>New technology</u> can have <u>risks</u>.

> For example, some scientists think using a mobile phone a lot may be <u>harmful</u>.

3) To make a <u>decision</u> about doing something that involves a <u>risk</u>, we need to think about:
   - the <u>CHANCE</u> of something bad happening,
   - how <u>SERIOUS</u> the <u>consequences</u> (results) would be if it did happen.

People Make Their Own Decisions About Risk

1) Not all risks have the same <u>consequences</u>.

> For example:
> If you <u>chop vegetables</u> with a sharp knife you risk <u>cutting your finger</u>.
> If you go <u>scuba-diving</u> you risk <u>death</u> — so the <u>consequences</u> are much more <u>serious</u>.

2) People are more likely to accept a risk if the consequences <u>don't last long</u> and <u>aren't serious</u>.
3) People are also more likely to accept a risk if they're <u>CHOOSING</u> to do something.
4) People are <u>less likely</u> to accept a risk if it's <u>FORCED</u> on them (for example, having a nuclear power station built next door).

The Benefits and Risks Need Weighing Up

1) New technologies have different <u>BENEFITS</u> and <u>RISKS</u> for <u>different people</u>.
2) For example, in building a <u>new nuclear power station</u>:

<u>Everyone gets electricity</u>.

<u>Builders</u> get work building the power station.

<u>Local people</u> get new jobs.

> BUT the <u>risks</u> are:
> - People living <u>nearby</u> might be exposed to <u>radiation</u>.
> - A <u>large area</u> might be in danger if there's a big <u>accident</u>.

3) To make a <u>decision</u> about whether to do something we need to <u>WEIGH UP</u> the benefits and risks involved for everyone.
4) <u>Governments</u> often have to decide what risks it's OK to take.

Ideas About Science

Science and Ethics

There are often issues to do with ETHICS in science — this means whether it's right or wrong to do something.

Some Questions Can't be Answered by Science

1) Experiments CAN'T EVER answer the question of whether something is ETHICALLY right or wrong.
2) That's because there is no right or wrong answer.
3) Take embryo screening (which means you can choose to have a child with particular features):

Some people say it's GOOD...	Some people say it's BAD...
For example, couples whose child needs a bone marrow transplant will be able to have another child with matching bone marrow. The new child can then donate bone marrow to the first child. This would save the life of their first child.	For example, they say it could have serious effects on the new child. The new child might feel that they were only wanted for their bone marrow.

4) The best we can do is make a decision that MOST PEOPLE are happy with.

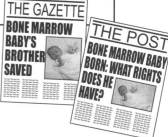

There Are Two Key Arguments About Ethical Problems

1
- Some people think that certain actions are ALWAYS wrong.
- They feel these actions are unacceptable, even if there are some benefits.

2
Some people say that the right decision is the one that benefits the most people.

The Law is Sometimes Involved Too

The LAW is involved in regulating (controlling) some areas of science. For example:

1) Animal research.
2) Genetic experiments.
3) Nuclear research.

Ideas About Science

Using Equations

Sometimes in science you have to do some maths. Boo. Hiss.
But if you learn how to use equations, they're a great way to pick up marks in the exam.

Most of the Equations You Need Are Written in the Exam Paper

1) Equations can LOOK tricky.

2) But for most equations, all you have to do is TIMES OR DIVIDE ONE NUMBER BY ANOTHER.

3) It's really useful to know equations off by heart.

4) But MOST of the equations you need will be on a special PAGE in the exam paper. Hooray.

5) You just have to know WHICH EQUATION to use and HOW to use it.

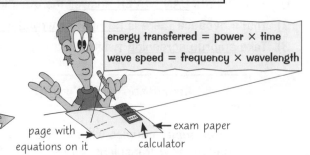

energy transferred = power × time
wave speed = frequency × wavelength

page with equations on it — exam paper — calculator

Example: A wave has a frequency of 20 Hz and a wavelength of 5 m. What is its speed?

① Decide WHICH EQUATION to use and WRITE IT OUT. → If you can't remember the equation, look for one on the equations page with frequency, wavelength and speed in it. → wave speed = frequency × wavelength

② PLUG IN THE NUMBERS. Sometimes you'll need to get them in the right units first — see below. → Frequency is 20 Hz, and wavelength is 5 m. Write these numbers under the equation. → wave speed = frequency × wavelength
wave speed = 20 Hz × 5 m

③ WORK OUT the answer with a calculator. → wave speed = 20 Hz × 5 m = 100

④ Don't forget the UNITS. → The units of speed are m/s. → wave speed = 100 m/s

---

Check Your Units

1) Before you plug the numbers in, check the numbers in the question have the RIGHT UNITS.

2) You need to LEARN what the RIGHT UNITS are for the things in the equations.

3) For example, wavelength ALWAYS needs to be in metres (m) to use the wave speed equation.

4) If you're given a wavelength in centimetres, you have to change it to metres BEFORE you use the equation.

Another Example: Find the energy transferred, in kWh, by a 1.5 kW hair drier in 30 minutes.

- Work out which equation you need to use: → energy transferred = power × time
- The power is 1.5 kW. The time is 30 minutes.
- But to get energy in kWh (kilowatt-hours), the time needs to be in HOURS.
- There are 60 minutes in an hour. So 30 minutes = 30 ÷ 60 = 0.5 hours.
- Now you can plug the numbers into the equation: energy transferred = power × time
energy transferred = 1.5 kW × 0.5 h
energy transferred = 0.75 kWh

5) Remember you always need to give the right units with your ANSWER too.

Using Equations

Module B1 — You and Your Genes

Genes, Chromosomes and DNA

Welcome to the first Biology bit of the book. You're going to love it.

1) Cells in your body have a NUCLEUS.

2) CHROMOSOMES are found in the nucleus.

3) Each chromosome is one very long bit of DNA.

4) A GENE is a short bit of DNA.

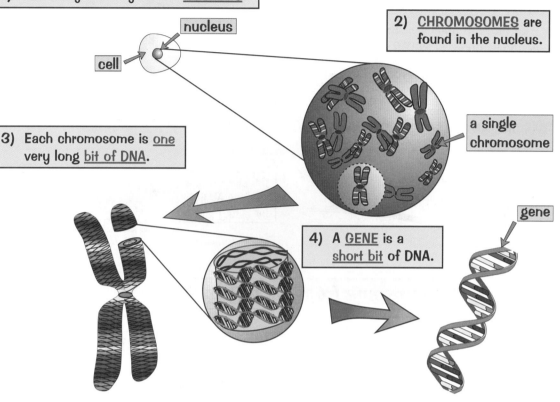

5) Genes control two things:
   - how an organism DEVELOPS (how it grows up).
   - how an organism FUNCTIONS (how it works).

'Organism' is just a posh word for 'living thing'.

Genes are Instructions for Proteins

A gene tells a cell how to make a PROTEIN. There are two types of proteins:

- STRUCTURAL PROTEINS (like collagen) — used to BUILD THINGS in the organism.
- FUNCTIONAL PROTEINS (like enzymes) — used to DO THINGS in the organism.

Characteristics are the Features of an Organism

Some characteristics are controlled by:

ONLY GENES	MANY GENES	ONLY THE ENVIRONMENT	GENES AND THE ENVIRONMENT
			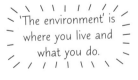
dimples	eye colour	scars	weight

 examples:

'The environment' is where you live and what you do.

Practice Questions

1) Where are chromosomes found?
2) Name one characteristic that's controlled by both genes and the environment.

Genes and Variation

That first page was just a little intro — here's where the fun stuff on genes really starts...

Body Cells Have Two of Each Chromosome

1) In body cells (like those in your skin or muscle) chromosomes come in PAIRS.

2) One chromosome in each pair has come from EACH PARENT.

3) BOTH chromosomes in a pair have the SAME GENES in the SAME PLACES.

4) The two chromosomes might have DIFFERENT VERSIONS of each gene.

5) DIFFERENT VERSIONS of the SAME GENE are called ALLELES.

Sex Cells Have One Chromosome from Each Pair

MALE sex cells are called SPERM.

FEMALE sex cells are called EGGS.

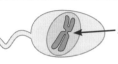

one chromosome from each pair (half the number of chromosomes)

The sex cells JOIN TOGETHER...

...during SEXUAL REPRODUCTION.

FERTILISED EGG

pairs of chromosomes (the full number of chromosomes)

SEXUAL REPRODUCTION makes VARIATION (differences) in OFFSPRING.

'Offspring' is another word for 'children'.

Children Look a Bit Like Both of Their Parents

Mum

Dad

Harriet

① Each child gets HALF of their alleles from their MUM...

② ...and HALF from their DAD.

③ This is why most children look a bit like BOTH of their parents.

④ This is Emily — Harriet's sister.

- Brothers and sisters DON'T look exactly THE SAME as each other.
- This is because they come from DIFFERENT sex cells with DIFFERENT alleles.

Practice Questions

1) What are different versions of the same gene called?
2) Do sex cells have one chromosome or both chromosomes from each pair?
3) "Children get all of their alleles from their mum." True or false?

Module B1 — You and Your Genes

Inheritance and Genetic Diagrams

Genetic diagrams help to work out how characteristics are passed on from parents to their children.

The Mixture of Alleles Controls the Characteristics

1) Scientists use LETTERS to show the ALLELES that an organism has.
2) You have TWO alleles for each gene. Each one can be either DOMINANT or RECESSIVE.
3) Big letters like 'D' are used for dominant alleles. Small letters like 'd' are used for recessive ones.

 If both alleles are DOMINANT, the DOMINANT characteristic will be shown.

If both alleles are RECESSIVE, the RECESSIVE characteristic will be shown.

 If you have one DOMINANT and one RECESSIVE allele only the DOMINANT characteristic will be shown.

Genetic Diagrams Show How Children Get Alleles From Their Parents

Genetic diagrams can look quite scary — but they're easier than they look.
The best way to learn them is with an example:

1) The pigs at Bacon Farm can be GREEN or NORMAL coloured.
2) Green colour is caused by the RECESSIVE allele 'n'.
3) Normal colour is caused by the DOMINANT allele 'N'.
4) The genetic diagram below shows what could happen when two NORMAL coloured pigs have a piglet. Both pigs have one DOMINANT allele and one RECESSIVE allele (Nn):

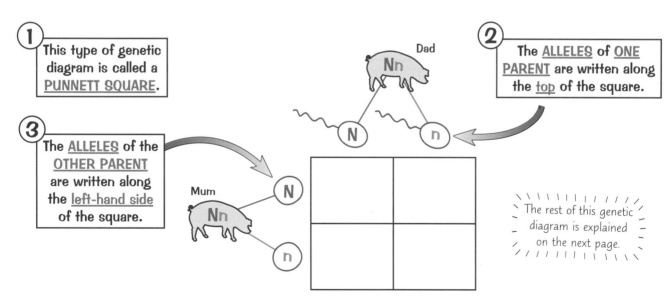

① This type of genetic diagram is called a PUNNETT SQUARE.

② The ALLELES of ONE PARENT are written along the top of the square.

③ The ALLELES of the OTHER PARENT are written along the left-hand side of the square.

The rest of this genetic diagram is explained on the next page.

Module B1 — You and Your Genes

Inheritance and Genetic Diagrams

4) The ALLELES are paired up in the boxes. The boxes show the different POSSIBLE alleles in a piglet.

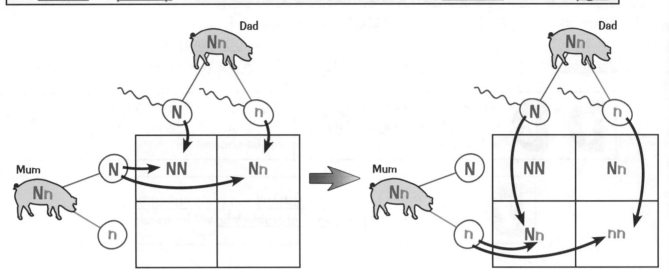

5) When you've filled in ALL the boxes, you can work out what a PIGLET could LOOK LIKE.

Remember — normal colour is the dominant characteristic.

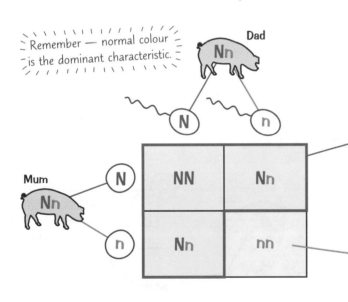

3 out of the 4 possible mixtures of alleles would make a NORMAL coloured piglet. There's a 75% CHANCE a piglet will be NORMAL coloured.

1 out of the 4 possible mixtures of alleles would make a GREEN piglet. There's a 25% CHANCE a piglet will be GREEN.

Practice Questions

1) Which characteristic is shown if one allele is dominant and one allele is recessive?

2) A green pig (nn) and a normal pig (Nn) from Bacon Farm have a piglet.
   a) Fill in the mum's alleles on the diagram to the right.
   b) Pair up the mum's and dad's alleles in the boxes.
   c) What is the percentage chance that these pigs will have a green piglet?

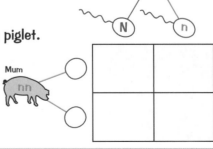

Module B1 — You and Your Genes

Genetic Diagrams and Sex Chromosomes

Just when you thought you'd finished with genetic diagrams these things called family trees show up.

Family Trees are Another Type of Genetic Diagram

FAMILY TREES also show how ALLELES are PASSED ON from parents to children.

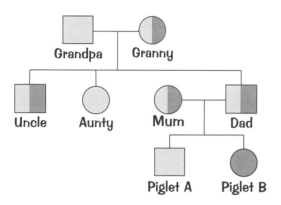

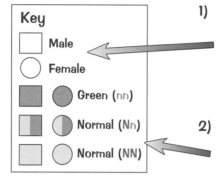

1) The SHAPES tell you whether the pig is a BOY or a GIRL. In this family tree boys are SQUARES and girls are CIRCLES.

2) The COLOUR of the shapes tells you which ALLELES each pig has (nn, Nn or NN).

Your Chromosomes Control Your Sex

1) You have SEX CHROMOSOMES that decide whether you're a BOY or a GIRL.

All BOYS have an X and a Y chromosome: XY

All GIRLS have two X chromosomes: XX

2) You can draw a genetic diagram to show how sex chromosomes are passed on to a child.

3) It just shows the sex chromosomes rather than different alleles.

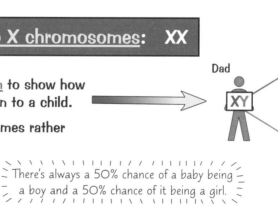

There's always a 50% chance of a baby being a boy and a 50% chance of it being a girl.

Practice Questions

1) Look at the family tree on the right.
   a) Is Alex a boy or a girl?
   b) Is Matt clumsy?
2) Which sex chromosomes do boys have?
3) Sam has two X chromosomes. Is Sam a boy or a girl?

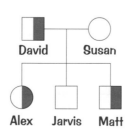

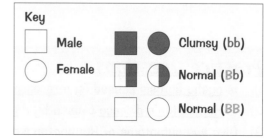

Module B1 — You and Your Genes

Genetic Disorders

The alleles of some genes can be faulty. This can cause some pretty nasty problems.
These problems are called genetic disorders — they can be passed on from parents to children.

Some Genetic Disorders Are Caused by Recessive Alleles

1) CYSTIC FIBROSIS is caused by a faulty RECESSIVE allele. We'll give this allele the letter 'f'.
2) The MIXTURE of alleles that a person has decides whether they have cystic fibrosis or not:

Alleles	Description of Person	Symptoms	Explanation
FF	NORMAL	NONE	People with TWO COPIES of the DOMINANT allele WON'T show any symptoms of cystic fibrosis.
Ff	CARRIER — someone with only one copy of the recessive allele, so isn't a sufferer.	NONE	CARRIERS WON'T show any symptoms. This is because they only have ONE COPY of the RECESSIVE allele.
ff	SUFFERER — someone who has the disorder.	• finds it hard to breathe • thick mucus (snotty gunk) • find it hard to digest food • chest infections (bad coughs)	Only people with TWO COPIES of the RECESSIVE allele are sufferers.

Symptoms are signs of disease.

Some Genetic Disorders Are Caused by Dominant Alleles

1) HUNTINGTON'S DISEASE is caused by a faulty DOMINANT allele. We'll give this allele the letter 'H'.
2) The MIXTURE of alleles that a person has decides whether they have Huntington's disease or not:

Alleles	Description of Person	Symptoms	Explanation
HH	SUFFERER	• tremors (shaking) • clumsy • memory loss • mood changes • can't concentrate	People with ONE or TWO COPIES of the DOMINANT allele have Huntington's disease.
Hh	SUFFERER		
hh	NORMAL	NONE	Only people with TWO COPIES of the RECESSIVE allele are normal.

3) Huntington's disease has a LATE ONSET. This means that sufferers only show symptoms later in their lives.

Practice Questions

1) Is cystic fibrosis caused by a recessive or dominant allele?
2) Is Huntington's disease caused by a recessive or dominant allele?
3) Give two symptoms of Huntington's disease.

Genetic Testing

Doctors and scientists can test for lots of different genetic disorders, but this isn't always a good thing.

Genetic Testing can be Really Useful

Genetic testing is just looking at a person's DNA to see what ALLELES they have.
There are three main uses of genetic testing:

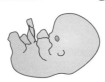

1) When doctors are helping a couple to have a baby, they can test the EMBRYOS for genetic disorders. Only healthy embryos are chosen to be put into the mother.

An embryo is a baby growing in the womb.

2) Couples can have tests before having children. They can find out if their children might inherit a genetic disorder.

3) People can be tested before they're given DRUGS. This can tell doctors if the drug will work well. It can also tell them if the person will have a DANGEROUS REACTION to it.

There are Issues With Genetic Testing

1) Genetic tests AREN'T ALWAYS RIGHT. This means that...

...HEALTHY PEOPLE could be told that they HAVE A GENETIC DISORDER. This is called a FALSE POSITIVE result.

...PEOPLE WITH A DISORDER could be told that they ARE FIT AND WELL. This is called a FALSE NEGATIVE result.

2) Tests carried out during pregnancy AREN'T ALWAYS SAFE. Collecting cells from an unborn baby sometimes causes a miscarriage (the baby dies).

3) If a test result is POSITIVE, there are lots of important questions to answer:

 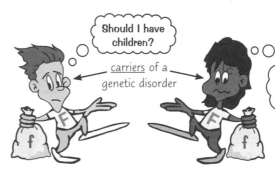

Do I still want to have my baby?

Should I have children?

carriers of a genetic disorder

Should I tell my family that they might be carriers too? What if they don't want to know?

4) If EMPLOYERS and INSURANCE COMPANIES get the results it might cause problems.

My insurance company saw the results of my genetic test. Now I can't get life insurance.

I have a genetic disorder and I might need lots of time off sick in the future. Employers might not give me a job if they find out.

Practice Questions

1) "Genetic testing can't be used on embryos." True or false?
2) Give one way genetic tests on unborn babies can be dangerous?

Module B1 — You and Your Genes

Clones

Nature has been making clones for millions of years. Don't let that put you off though — it's still interesting...

Clones have Identical Genes

1) Clones have exactly the SAME GENES as each other.
2) Any DIFFERENCES between clones are down to the ENVIRONMENT.

EXAMPLE: If you ate more food than your clone, you would be bigger — even though you still had exactly the same genes.

Nature Makes Clones...

1. By Asexual Reproduction

There's only one parent in asexual reproduction.

Lots of organisms can form clones by reproducing ASEXUALLY (without sex):

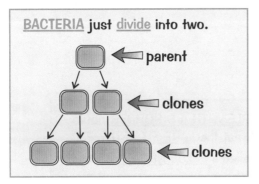

BACTERIA just divide into two.
← parent
← clones
← clones

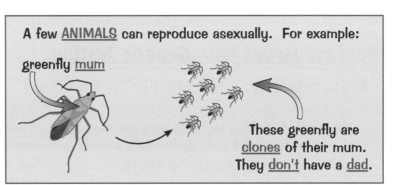

A few ANIMALS can reproduce asexually. For example:
greenfly mum

These greenfly are clones of their mum. They don't have a dad.

Some PLANTS grow stems called RUNNERS. Runners spread out and form CLONES at their tips.

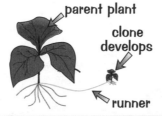

parent plant
clone develops
runner

Other PLANTS make underground BULBS. These can grow into CLONES.

2. When Cells of an Embryo Split

1) IDENTICAL TWINS are also CLONES.
2) Identical twins are formed when an EMBRYO splits into TWO.

Practice Questions

1) What causes differences between clones?
2) What type of reproduction produces clones — sexual or asexual?
3) "Identical twins are not clones." True or false?

Module B1 — You and Your Genes

Stem Cells

Stem cells are <u>pretty cool</u> — they can turn into <u>different types</u> of cells...

Stem Cells *Can Become Other Types* of Cells

1) Most <u>cells</u> in your body are <u>SPECIALISED</u> to do a certain task. For example, <u>red blood cells</u> carry oxygen.
2) Most cells in an organism become <u>SPECIALISED</u> when the organism is really young.
3) Some cells are <u>UNSPECIALISED</u>.
4) <u>UNSPECIALISED</u> cells are called <u>STEM CELLS</u>. They can develop into <u>different types of cells</u>.

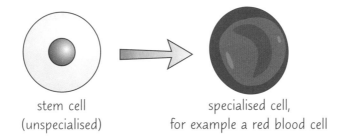

stem cell
(unspecialised)

specialised cell,
for example a red blood cell

There are *TWO Main Types* of Stem Cell

TYPE	WHERE ARE THEY FOUND?	WHAT TYPES OF CELL CAN THEY TURN INTO?
<u>EMBRYONIC STEM CELLS</u>	<u>EMBRYOS</u>	<u>ANY</u> type of cell
<u>ADULT STEM CELLS</u>	<u>ADULTS</u>	<u>MANY</u> (but not all) types of cell

Stem Cells *Could be Used to Help Ill People*

1) <u>STEM CELLS</u> could be used by doctors to <u>treat SICK PEOPLE</u>.
2) Stem cells could be turned into <u>NEW CELLS</u> that can <u>REPLACE DAMAGED</u> cells. For example, new heart cells for people with heart disease.

Practice Questions

1) Are stem cells <u>specialised</u> or <u>unspecialised</u>?
2) Name <u>two places</u> where you can find stem cells.
3) What type of cell can <u>embryonic stem cells</u> turn into?

Module B1 — You and Your Genes

Module B2 — Keeping Healthy

Microorganisms and Disease

Microorganisms can make you ill. And as if that wasn't bad enough, they reproduce (copy themselves) inside you too. Eww.

Microorganisms Cause Disease

1) Microorganisms are things like BACTERIA and VIRUSES.
2) Some of them cause infectious diseases.
3) Infectious diseases often have SYMPTOMS, for example a rash.
4) Microorganisms cause symptoms by:

Infectious diseases are diseases that can be caught.

① DAMAGING YOUR CELLS

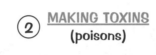
② MAKING TOXINS (poisons)

Microorganisms Reproduce Quickly In Humans

1) Microorganisms reproduce (copy themselves) really quickly inside the human body.
2) This is because the body has the right conditions:
   - It's WARM.
   - It's MOIST.
   - There's FOOD.

It doesn't get any better than this...

You can Work Out How Big a Group of Microorganisms Will Grow

1) When one microorganism copies itself you get two microorganisms.
2) The number of microorganisms in a group will DOUBLE each time the microorganisms copy themselves:

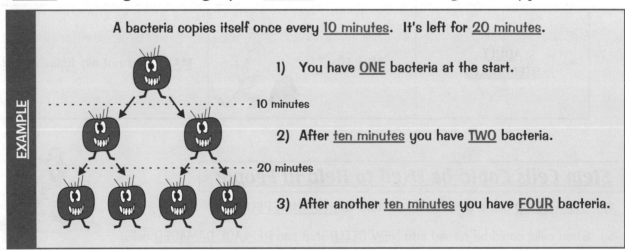

EXAMPLE

A bacteria copies itself once every 10 minutes. It's left for 20 minutes.

1) You have ONE bacteria at the start.
- - - 10 minutes
2) After ten minutes you have TWO bacteria.
- - - 20 minutes
3) After another ten minutes you have FOUR bacteria.

Practice Questions

1) Give two ways that microorganisms can cause symptoms.
2) Give three conditions in the human body that help microorganisms reproduce quickly.
3) You start with one bacteria that copies itself every 30 minutes. How many will you have after 90 minutes?

The Immune System

White blood cells deal with microorganisms that get into your body. You need to know how they do this.

Your Immune System Fights Off Microorganisms

1) Your immune system kills microorganisms that get into your body.
2) The most important bit of the immune system is the WHITE BLOOD CELLS. They do two things:

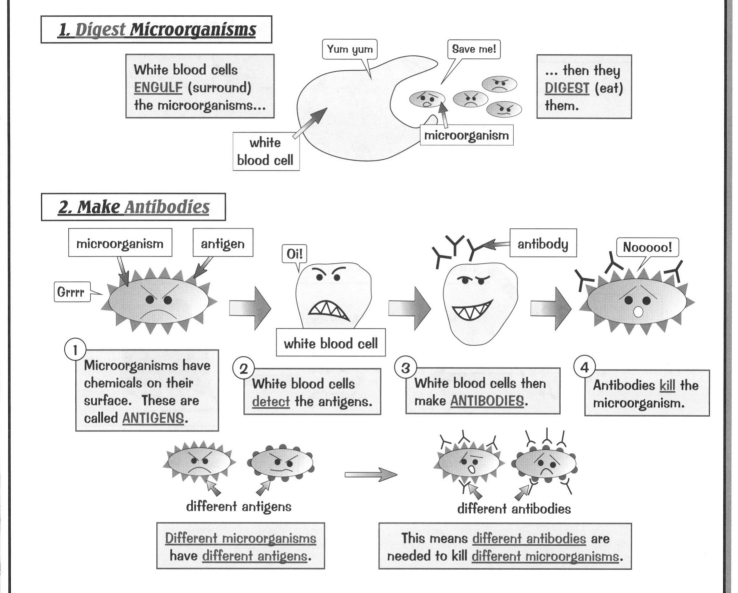

Memory Cells Stop You From Getting the Same Illness Again

1) There are special white blood cells called MEMORY CELLS.
2) These remember how to make certain antibodies.
3) Memory cells make the antibodies really quickly if a microorganism SHOWS UP AGAIN.
4) This stops a person from getting the same illness again. It gives them IMMUNITY.

Practice Questions

1) Give two ways that white blood cells kill microorganisms.
2) What are antigens?
3) "Memory cells remember how to make certain antigens." True or false?

Module B2 — Keeping Healthy

Vaccination

Some diseases are so horrible you really don't want to get them. But don't worry — you can help your body fight the little nasties that cause them. You do this by having a vaccination. Oh what fun...

Vaccinations Are Injections of Dead Microorganisms

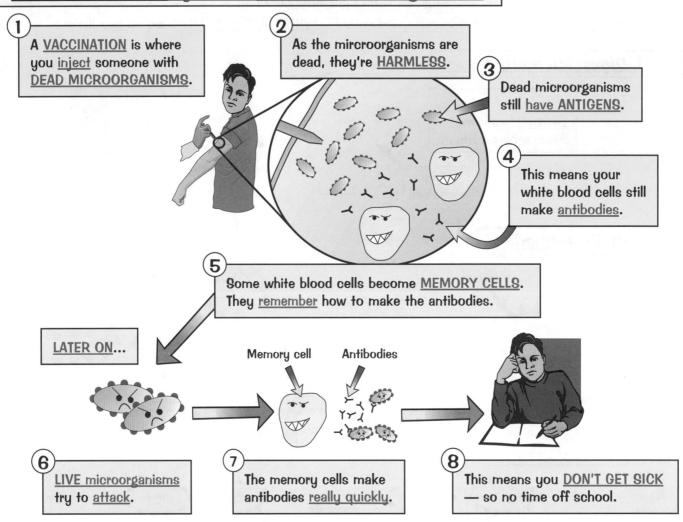

1) A VACCINATION is where you inject someone with DEAD MICROORGANISMS.
2) As the mircroorganisms are dead, they're HARMLESS.
3) Dead microorganisms still have ANTIGENS.
4) This means your white blood cells still make antibodies.
5) Some white blood cells become MEMORY CELLS. They remember how to make the antibodies.

LATER ON...

6) LIVE microorganisms try to attack.
7) The memory cells make antibodies really quickly.
8) This means you DON'T GET SICK — so no time off school.

Vaccinations and Drugs have Different Effects on Different People

1) Vaccinations and drugs (medicines) can have SIDE EFFECTS. For example, they can make people throw up.
2) The side effects can be WORSE for SOME people than for others.
3) One reason people react differently is because of DIFFERENCES in their GENES.

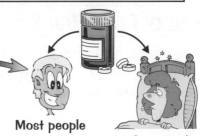

Most people   A few people

Practice Questions

1) "Dead microorganisms don't have any antigens." True or false
2) Why do people react differently to vaccinations and drugs?

Antibiotics

Antibiotics are good at killing those pesky bacteria. But they're not perfect...

Antibiotics Can Kill Bacteria

Antibiotics are a type of antimicrobial (chemicals that kill microorganisms).

Bacteria Can Become Resistant to Antibiotics

1) Over time bacteria can change. For example, bacteria can become RESISTANT to antibiotics.
2) This means antibiotics DON'T AFFECT them.
3) You can make it HARDER for bacteria to become resistant to antibiotics by:

1 Taking ALL the antibiotics a doctor gives you.

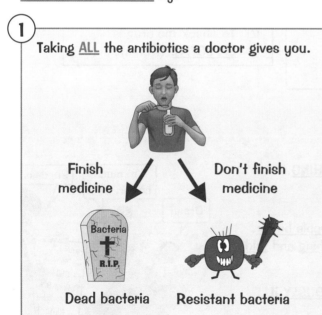

2 Only using antibiotics when you REALLY NEED TO. For example, for serious infections and not for things like sore throats.

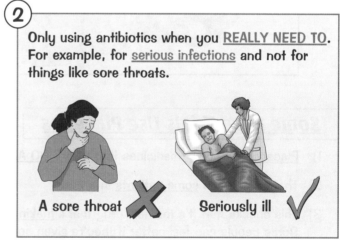

A sore throat ✗   Seriously ill ✓

Practice Questions

1) Name one type of microorganism antibiotics can kill.
2) Name one type of microorganism antibiotics can't kill.
3) Give two things you can do to make it harder for bacteria to become resistant to antibiotics.

Module B2 — Keeping Healthy

Drug Trials

Any drug has to be tested before it's used. This is to make sure it's safe and does what it's supposed to do.

Drugs Go Through Three Stages of Testing

1) LABORATORY TESTING

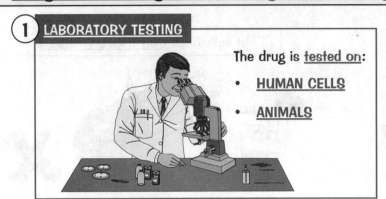

The drug is tested on:
- HUMAN CELLS
- ANIMALS

REASONS
1) To check the drug is SAFE.
2) To check the drug is EFFECTIVE (works).

2) TESTING ON HEALTHY VOLUNTEERS

REASON
To check the drug is SAFE.

3) TESTING ON ILL VOLUNTEERS

REASONS
1) To check the drug is SAFE.
2) To check the drug is EFFECTIVE (works).

Some Drug Trials Use Placebos

1) Placebos are 'fake' medicines that DON'T DO ANYTHING.

2) They're given to some patients in a trial.

3) This checks that it's the new drug that's making people better. Some people can feel better if they're given something and told it'll help them.

4) Placebos are NOT GIVEN to patients who are SERIOUSLY ILL.

5) This is because it's NOT FAIR to give seriously ill patients a fake medicine if the new drug could help them.

Practice Questions

1) Name two ways that drugs are tested in a laboratory.
2) Why are drugs tested on a) healthy volunteers, b) ill volunteers?
3) Why aren't placebos given to patients who are seriously ill?

Module B2 — Keeping Healthy

The Circulatory System

The circulatory system is all the bits and bobs that are involved in getting blood around your body.

The Heart and Blood Vessels Supply Blood to the Body

1) Blood travels around the body in tubes called BLOOD VESSELS.
2) The heart pumps blood through the vessels.
3) The heart is a DOUBLE PUMP.
   This means it's two pumps stuck together.
   - The RIGHT SIDE pumps blood to the LUNGS.
   - The LEFT SIDE pumps blood around the rest of the BODY.
4) The heart's made up of MUSCLE CELLS.
5) The heart muscle cells have their own blood supply.
6) The blood brings them stuff like FOOD and OXYGEN to stay alive.
7) This keeps the heart beating all the time.

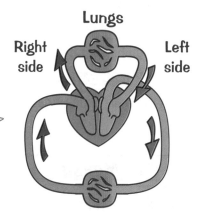

There are Three Types of Blood Vessel

Blood vessel	Function	Structure
ARTERIES	Carry blood AWAY FROM the heart at HIGH PRESSURE.	THICK, STRETCHY WALLS — to cope with the HIGH PRESSURE.
CAPILLARIES	Carry blood TO tissues (groups of cells).	THIN WALLS — so things like food and oxygen can PASS THROUGH EASILY.
VEINS	Carry blood TO the heart at LOW PRESSURE.	LARGE HOLE IN MIDDLE — this helps blood flow easily. VALVES — these keep blood flowing in the RIGHT DIRECTION.

Practice Questions

1) The heart is a double pump. What does this mean?
2) Give one feature of arteries. Say how this feature helps them carry blood around the body.
3) Give one feature of veins. Say how this feature helps them carry blood around the body.

Module B2 — Keeping Healthy

Heart Rate and Blood Pressure

Your blood has got to travel a long way around your body. The only way it can do this is if it's under pressure. Your heart beats all the time to keep the blood moving and keep up this pressure.

Your Pulse Rate is The Same As Your Heart Rate

1) You can feel your PULSE on the inside of your wrist.
2) The number of pulses you feel in ONE MINUTE is your pulse rate.
3) Your PULSE RATE is the SAME as your HEART RATE.
4) So you can find out your HEART RATE by measuring your PULSE RATE.

You Can Measure Your Blood Pressure

1) BLOOD PRESSURE is how hard your blood is PUSHING against an ARTERY WALL. You can measure this.
2) A blood pressure measurement is given as TWO NUMBERS:

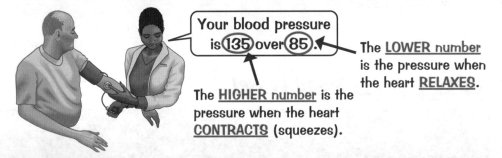

1) You can compare your heart rate and blood pressure to "NORMAL" MEASUREMENTS.
2) These are what the measurements should be when you're FIT and HEALTHY.
3) Normal measurements are given as RANGES (for example, between 80 and 100). This is because INDIVIDUAL PEOPLE are DIFFERENT.

High Blood Pressure Increases the Risk of Heart Disease

Heart attacks are a type of heart disease.

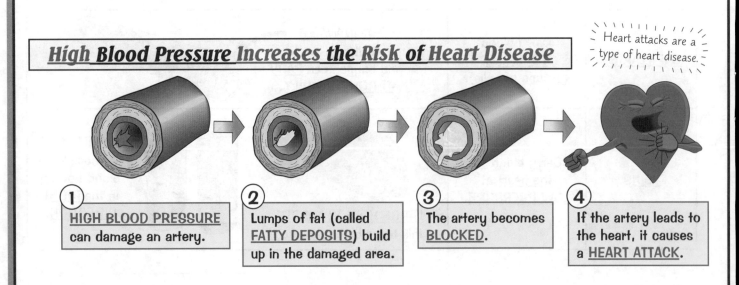

① HIGH BLOOD PRESSURE can damage an artery.
② Lumps of fat (called FATTY DEPOSITS) build up in the damaged area.
③ The artery becomes BLOCKED.
④ If the artery leads to the heart, it causes a HEART ATTACK.

Practice Questions

1) What is your pulse rate the same as?
2) What is blood pressure?
3) High blood pressure can damage arteries. How can this lead to a heart attack?

Module B2 — Keeping Healthy

Heart Disease

The heart is pretty important — but the things we do (our lifestyles) can damage it...

Lifestyle Factors Can Increase the Risk of Heart Disease

1) Here are the five main ones:

① POOR DIET   ② STRESS   ③ too much ALCOHOL   ④ SMOKING

⑤ ILLEGAL DRUGS like cannabis and ecstasy

- DRUGS like alcohol, nicotine, ecstasy and cannabis can increase your BLOOD PRESSURE and HEART RATE.
- This can increase your risk of a HEART ATTACK.

Nicotine is a drug in cigarettes.

2) Some people might be more at risk of heart disease because of their GENES too.
3) REGULAR EXERCISE lowers the risk of heart disease.
4) Heat disease is MORE COMMON in RICHER COUNTRIES than in poorer countries. This is because...

In RICHER COUNTRIES:
People have a HIGH-FAT DIET.
People DON'T do lots of EXERCISE.

In POORER COUNTRIES:
People have a LOW-FAT DIET.
People have to WALK MORE.

Studies Can Find Out what Increases the Risk of Heart Disease

1) Studies are used to FIND OUT the factors (things) that INCREASE THE RISK of heart disease.
2) For example, if a group of people died from heart disease scientists could see if they had:

SIMILAR LIFESTYLES, for example, did they all smoke?   SIMILAR GENES.

Practice Questions

1) Give three things that increase your risk of heart disease.
2) Give two reasons why heart disease is more common in richer countries than in poorer countries.
3) How could scientists find out what increases the risk of heart disease?

Module B2 — Keeping Healthy

Homeostasis and The Kidneys

Homeostasis — a word that strikes fear into the heart of many GCSE students. But it's really not that bad.

Homeostasis — Keeping the Conditions in your Body Steady

1) Homeostasis is all about BALANCING the stuff going into your body with the stuff leaving.

2) Homeostasis keeps the conditions inside your body STEADY. This means your cells can WORK PROPERLY.

3) Automatic control systems keep conditions steady. They're made up of THREE PARTS:

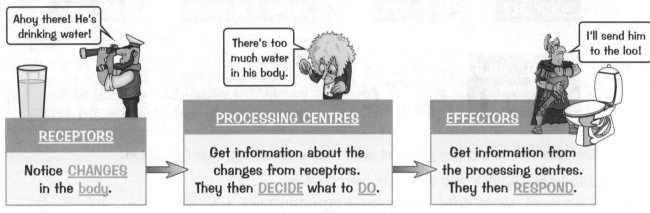

RECEPTORS — Notice CHANGES in the body.

PROCESSING CENTRES — Get information about the changes from receptors. They then DECIDE what to DO.

EFFECTORS — Get information from the processing centres. They then RESPOND.

4) Information is sent between receptors, processing centres and effectors by:
  - NERVES.
  - HORMONES (chemicals in the body).

Kidneys Help Balance Water and Other Things in the Body

The kidneys BALANCE the levels of:
1) WATER and
2) WASTE.

Balancing Water Level is Really Important

1) Your body needs to keep the water level in the cells just right so they work properly.

2) This means that your body needs to balance the water coming in and going out...

Water COMES IN from DRINK, FOOD
Water LEAVES by SWEATING, BREATHING, WEE, POO

Practice Questions

1) "Homeostasis keeps the conditions inside your body steady." True or false?
2) Name the three parts of automatic control systems.
3) a) Give two ways that water comes into the body.
   b) Give two ways that water leaves the body.

Module B2 — Keeping Healthy

Controlling Water Content

Controlling the amount of water in your body is pretty important — so here's a page about urine. Just for you.

Your Urine isn't Always the Same

The KIDNEYS balance water levels by making:

DILUTE URINE OR CONCENTRATED URINE
— contains lots of water   — contains a little water

The AMOUNT OF WATER in the URINE depends on the AMOUNT OF WATER in the BLOOD. This changes with:

1) OUTSIDE TEMPERATURE

1) For example, if it's hot outside, you get hot so you SWEAT. Sweat contains water. So when you sweat your body LOSES WATER.
2) This means there's LESS WATER in your BLOOD.
3) This means you make CONCENTRATED URINE.

2) EXERCISE

1) When you exercise, you get hot and you SWEAT.
2) Your body LOSES WATER. This means there's LESS WATER in your BLOOD.
3) This means you make CONCENTRATED URINE.

3) DRINKING FLUIDS AND EATING SALT

1) If you drink a lot you make DILUTE URINE.
2) But if you drink too little you make CONCENTRATED URINE.
3) Eating too much SALT also means you make CONCENTRATED URINE.

Alcohol and Ecstasy Affect Urine Production

Drinking Alcohol Leads to More Dilute Urine

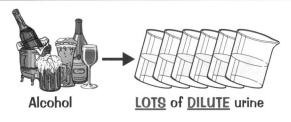

Alcohol   LOTS of DILUTE urine

1) Making LOTS of DILUTE URINE can cause DEHYDRATION.
2) This is where you don't have enough water in your body.
3) Dehydration can cause headaches, dizziness and even death.

Taking Ecstasy Leads to More Concentrated Urine

Ecstasy   A SMALL AMOUNT of CONCENTRATED urine

Practice Questions

1) How do the kidneys balance water levels?
2) Name two things that change the amount of water in your blood.
3) Does alcohol make you produce more dilute or more concentrated urine?

Module B2 — Keeping Healthy

Module B3 — Life on Earth

Adaptation and Variation

Life on Earth — I can't imagine what we'd do without it... well, we wouldn't exist at all I suppose.

Learn This Definition of a Species First...

A **SPECIES** is a **GROUP** of organisms that can **BREED** together to make **FERTILE OFFSPRING**.

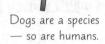 Dogs are a species — so are humans.

'Offspring' is another word for children. 'Fertile' just means able to have children.

Species are Adapted to Their Environments

1) If a species is well ADAPTED to its environment, it means it's got FEATURES that help it to SURVIVE there. For example, the CACTUS is adapted to living in DESERTS:

2) Members of a species are MORE LIKELY to have OFFSPRING if the species is well ADAPTED to its environment.

3) If a SPECIES has lots of offspring, the species is more likely to SURVIVE.

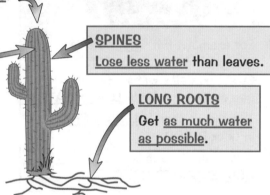

SPINES Lose less water than leaves.

THICK STEM Stores water for when there's not much around.

LONG ROOTS Get as much water as possible.

Individuals of the Same Species Have Differences

An 'individual' is just a single thing.

1) Individuals of a species are usually **DIFFERENT** — there's **VARIATION** between them.

2) Variation can be **GENETIC**. It's caused by **CHANGES IN GENES**. These changes are called **MUTATIONS**.

3) Mutations in **SEX CELLS** (sperm and egg cells) can be **PASSED ON** to offspring.

4) This can cause offspring to develop **NEW FEATURES**.

Practice Questions

1) What is a species?
2) Give three ways in which a cactus is adapted to where it lives.
3) a) What are mutations?
   b) What can genetic variation cause offspring to develop?

Natural Selection and Selective Breeding

A species becomes <u>better adapted</u> to where it lives by <u>natural selection</u>.

Natural Selection *is How* Species Adapt

Here's how <u>NATURAL SELECTION</u> works:

EXAMPLE

① Living things <u>VARY</u>.

<u>Some</u> rabbits have <u>big ears</u> and <u>some</u> have <u>small ears</u>.

② Some individuals have <u>FEATURES</u> that make them <u>more likely</u> to <u>SURVIVE</u>.

<u>Big-eared</u> rabbits are <u>more likely</u> to hear a fox sneaking up on them than <u>small-eared</u> rabbits.

③ The individuals with these features are more likely to <u>HAVE OFFSPRING</u>. This means they'll pass on the <u>GENES</u> that control these features.

This means they're <u>more likely</u> to <u>live</u> and <u>have babies</u> with <u>big ears</u>.

④ This means that the <u>features</u> which help individuals <u>survive</u> become <u>MORE COMMON</u>.

So pretty soon there'll be <u>more big-eared rabbits</u> than short-eared rabbits.

Selective Breeding *is Where Humans* Choose *The Features*

In selective breeding humans <u>CHOOSE</u> a <u>feature</u> they <u>want to appear</u> in offspring.

They <u>only breed</u> the organisms <u>with that feature</u>.

This means all the <u>OFFSPRING</u> will <u>have the feature too</u>.

<u>Lots</u> of milk   <u>Not a lot</u> of milk

1) Selective breeding is <u>DIFFERENT</u> from <u>natural selection</u>.
2) In natural selection <u>only</u> features that <u>HELP SURVIVAL</u> get passed on.

Practice Questions

1) "In natural selection, <u>features</u> that help <u>survival</u> become <u>more common</u>." True or false?
2) How is natural selection <u>different</u> from selective breeding?

Module B3 — Life on Earth

Evolution

There are lots of species around these days and evolution is how they all came about.

Somehow, a Long Time Ago, Life Must Have Started

1) LIFE on Earth began about 3500 MILLION YEARS AGO.
2) Today, there are LOADS of species on Earth.
3) Loads of other species have also become EXTINCT (they've died out).
4) All species that have ever lived EVOLVED from VERY SIMPLE living things.
5) EVOLUTION is the CHANGE IN A SPECIES over TIME.

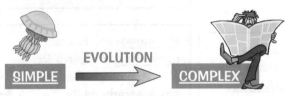

Evolution Can Make New Species

1) Groups of organisms of the same species can become ISOLATED (separated).
2) This means they CAN'T BREED together.

3) If the ENVIRONMENT CHANGES, the groups will develop different features.
4) MUTATIONS make the new features.
5) Some new features help the organism SURVIVE.
6) NATURAL SELECTION (see page 29) makes these features more common.

7) After a long time, these DIFFERENT GROUPS become DIFFERENT SPECIES.

Practice Questions

1) When did life begin on Earth?
2) Have all species evolved from complex things or from simple things?
3) "Evolution can make new species." True or false?

Module B3 — Life on Earth

Evolution

There's loads of evidence for evolution. But there was more than one theory of how evolution happens...

There is Good Evidence for Evolution

① THE FOSSIL RECORD

1) Some DEAD ORGANISMS are found in rocks. These are called FOSSILS.
2) The FOSSIL RECORD puts fossils in date order from oldest to youngest.
3) The fossil record shows organisms getting MORE COMPLEX — it SHOWS EVOLUTION.

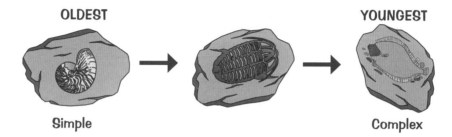

② DNA

1) All living things have SIMILAR DNA.
2) Species that are CLOSELY RELATED have VERY SIMILAR DNA.

3) Species that are LESS CLOSELY RELATED have LESS SIMILAR DNA.

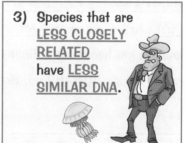

4) This means scientists can look at DNA to work out how life EVOLVED.

Evolution by Natural Selection Was a Clever Idea

1) CHARLES DARWIN came up with the theory of EVOLUTION by NATURAL SELECTION (see page 29).
2) But a French man called LAMARCK had a different idea. He thought that:

- If a feature was used a lot, it would become MORE DEVELOPED. For example, if you work out you get bigger muscles.
- This developed feature would be PASSED ON to offspring.

3) But developed features AREN'T caused by GENES.
4) This means they CAN'T be passed on to your children.
5) So Lamarck's theory was WRONG.

Practice Questions

1) Give two pieces of evidence for evolution.
2) Lamarck thought developed features were passed on to offspring. Why is this wrong?

Module B3 — Life on Earth

Biodiversity and Classification

Biodiversity is all the variety of life on Earth. Classification puts organisms into groups.

Earth's Biodiversity is Important

1) Biodiversity is the VARIETY OF LIFE on Earth. It includes:

The NUMBER of species.

The RANGE of different types of organisms, for example, from microorganisms to plants and animals.

The GENETIC VARIATION in a species (see page 28).

2) Biodiversity FALLS when species become EXTINCT (see below).

3) We need to stop extinction because biodiversity helps us find new CROPS and new MEDICINES.

The Speed of Extinction of Species is Increasing

1) A species is EXTINCT when there are NO MORE left.
2) The SPEED that animals are becoming extinct is GETTING FASTER.
3) This is mostly down to things HUMANS do, like hunting.

Classification is all About Putting Things into Groups

1) Scientists group organisms together based on how SIMILAR their DNA is.
2) They also group organisms on how SIMILAR their PHYSICAL FEATURES are. For example:

All VERTEBRATES have a skeleton with a BACKBONE.

All FLOWERING PLANTS have FLOWERS.

3) All living things are put into groups called KINGDOMS. Each kingdom is split up into more groups until you get down to a SPECIES:

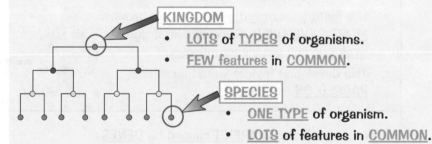

KINGDOM
- LOTS of TYPES of organisms.
- FEW features in COMMON.

SPECIES
- ONE TYPE of organism.
- LOTS of features in COMMON.

4) Putting organisms into groups shows us which species they EVOLVED FROM.

Practice Questions

1) What is biodiversity?
2) Give one thing that scientists use to group organisms together.
3) Give one difference between a kingdom and a species.

Module B3 — Life on Earth

Energy in an Ecosystem

An <u>ecosystem</u> is all the <u>different organisms</u> living together in a <u>certain place</u>. Sounds cosy.

A Food Chain Shows What is Eaten By What in an Ecosystem

Take a look at this <u>food chain</u>:

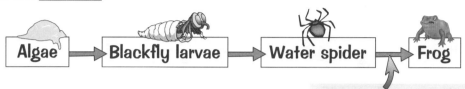

Algae → Blackfly larvae → Water spider → Frog

1) The chain is split up into <u>STAGES</u>.
2) Each stage is <u>an organism</u> in the <u>food chain</u>.
3) So in the food chain above, the <u>first stage</u> is the <u>algae</u>.
4) The <u>second stage</u> is the <u>blackfly larvae</u>, and so on.

The <u>ARROWS</u> show <u>what is eaten by what</u>. For example, the <u>water spider</u> is eaten by the <u>frog</u>.

The Energy in an Ecosystem Comes From the Sun

1) Energy comes from the <u>SUN</u>.

2) <u>PLANTS</u> use some of the Sun's energy during <u>PHOTOSYNTHESIS</u>.

3) Photosynthesis turns the energy into <u>chemicals</u>. The chemicals are <u>STORED</u> in the plants.

4) Energy is <u>PASSED</u> between organisms when:
   - animals <u>EAT</u> plants and other animals.
   - <u>decay organisms</u> (called decomposers) feed on <u>dead organisms</u> and <u>waste products</u>.

5) Energy is <u>lost</u> at each stage of the food chain as:
   - <u>HEAT</u>
   - <u>WASTE PRODUCTS</u> (for example, droppings and urine)
   - <u>UNEATEN PARTS</u> (for example, bones)

You <u>DON'T</u> normally get <u>food chains</u> with more than <u>FIVE STAGES</u>. This is because so much <u>ENERGY</u> is <u>LOST</u> at each stage.

Practice Questions

1) Look at the <u>food chain</u> at the top of the page. What do <u>water spiders</u> eat?
2) <u>Where</u> does the <u>energy</u> in an ecosystem come from?
3) Why don't food chains normally have <u>more</u> than <u>five</u> stages?

Module B3 — Life on Earth

Energy in an Ecosystem

I'm not going to lie to you — that is maths you can see on the page. But it's not as bad as it looks. Promise.

You Need to be Able to Use Data on Energy Flow

1) Take a look at this food chain:

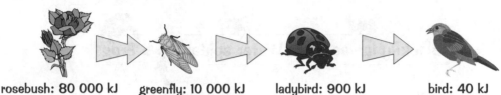

rosebush: 80 000 kJ   greenfly: 10 000 kJ   ladybird: 900 kJ   bird: 40 kJ

2) The numbers show the AMOUNT OF ENERGY in each stage. For example, there's 80 000 kJ in the rosebush.

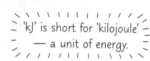

'kJ' is short for 'kilojoule' — a unit of energy.

3) Not all of the energy gets to the next stage though — most of it is LOST.

You Can Work Out How Much Energy is Lost at Each Stage

You just TAKE AWAY the energy in one stage from the energy in the previous stage. Like this:

Energy lost between the FIRST STAGE and the SECOND STAGE. = 80 000 kJ (rosebush) — 10 000 kJ (greenfly) = 70 000 kJ

You Can Work Out the Efficiency of Energy Transfer in a Food Chain

1) The EFFICIENCY OF ENERGY TRANSFER just means HOW WELL ENERGY PASSES from one stage to the next stage.

2) You can work it out using this FORMULA:

$$\text{efficiency} = \frac{\text{energy in one stage}}{\text{energy in the previous stage}} \times 100$$

Efficiency is given as a percentage (%). The lower the percentage, the less efficient the transfer is.

3) You just have to put the numbers in. For example:

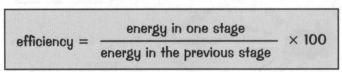

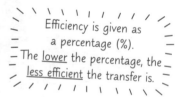

Efficiency of transfer from the FIRST STAGE to the SECOND STAGE. = $\frac{10\,000 \text{ kJ}}{80\,000 \text{ kJ}} \times 100$ = 12.5% efficient

Practice Questions

1) Some shrimp contain 10 000 kJ of energy. They're eaten by a fish, which contains 1000 kJ. Work out:
   a) The amount of energy lost between the stages.
   b) The efficiency of energy transfer.

Use the formula above to help you.

Module B3 — Life on Earth

Interactions Between Organisms

Resources are things that organisms need from their environment and other organisms to survive. For example, food and water.

Living Things Compete for Resources

Organisms get the things they need to SURVIVE (like water) from their ENVIRONMENT...

... but resources are LIMITED (there aren't many of them).

So different species have to COMPETE for resources...

... but some organisms won't get enough — so they WON'T SURVIVE.

Any Change in Any Environment can Have Knock-on Effects...

1) The diagram on the right shows a FOOD WEB. A food web shows what is eaten by what.

2) Organisms DEPEND on other species for their SURVIVAL. For example, they need other animals for food.

- Imagine all the WATER SPIDERS were killed.

- The FROGS would have nothing to eat. The number of frogs would DECREASE.

- But there would be nothing to eat the STONEFLY and BLACKFLY LARVAE. Their numbers would INCREASE.

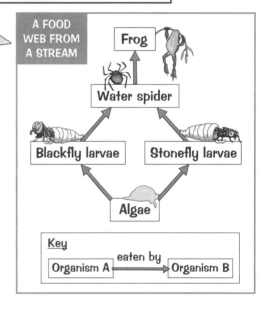

...Even Extinction

A species could be made EXTINCT if:

① The ENVIRONMENT CHANGES. For example, a species' home is destroyed and it can't cope.

② Something new turns up. For example:
- A PREDATOR (something that hunts the species).
- Something that COMPETES with the species for resources.
- Something that causes DISEASE.

Remember — a species is extinct when there are no more left.

③ An organism that the species eats becomes EXTINCT. This can mean the species doesn't have enough to eat.

Practice Questions

1) What does a food web show?
2) Give one change that can make a species extinct.

Module B3 — Life on Earth

The Carbon Cycle

Lots of people don't like the carbon cycle, but I happen to love it. I hope you will too...

The Carbon Cycle Shows How Carbon is Recycled

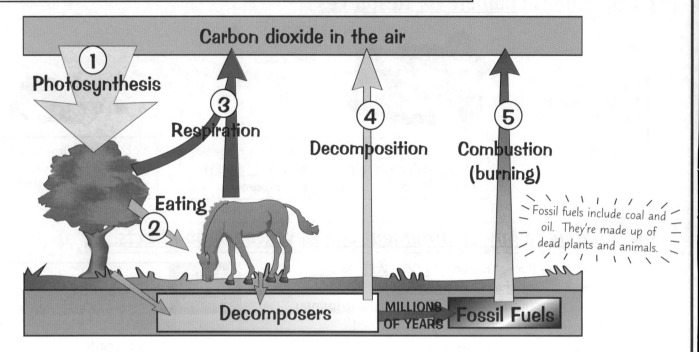

You need to learn these important points:

1) PLANTS turn the CARBON from carbon dioxide in the air into SUGARS. This is PHOTOSYNTHESIS.

2) EATING passes the sugars in the plants along to ANIMALS.

3) Plant and animal RESPIRATION releases carbon dioxide back into the air.

4) Plants and animals eventually die and DECOMPOSE. This is when they're BROKEN DOWN by microorganisms called DECOMPOSERS. Decomposers release carbon dioxide into the air.

Microorganisms are tiny little creatures like bacteria.

5) The COMBUSTION (burning) of fossil fuels also releases carbon dioxide into the air.

Practice Questions

1) What does the carbon cycle show?
2) How do plants turn carbon in the air into sugars?
3) Give three ways carbon is released into the air.

The Nitrogen Cycle

Nitrogen, just like carbon, is always being recycled. So the nitrogen plants and animals use might once have been in the air. And before that it might have been in a plant... or a nice, little hedgehog...

Nitrogen is Recycled in the Nitrogen Cycle

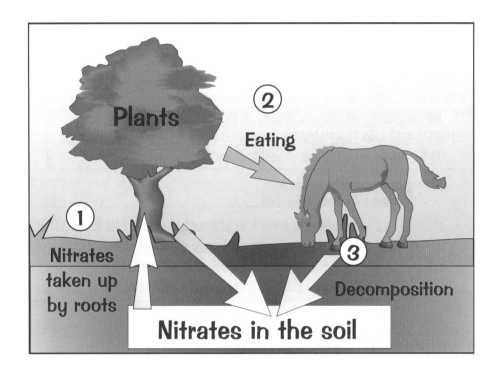

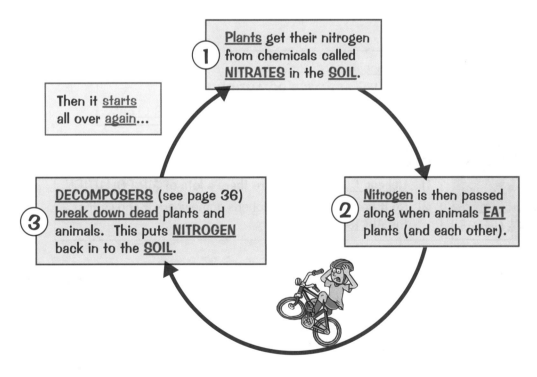

Practice Questions
1) Where do plants get nitrogen from?
2) Where do animals get nitrogen from?
3) Which organisms put nitrogen back into the soil?

Module B3 — Life on Earth

Measuring Environmental Change

Environments are changing all the time. Some of these changes can be measured.

Environmental Change can be Measured with Non-Living Indicators...

TEMPERATURE

1) You can measure the TEMPERATURE of an environment.
2) You can do this with a thermometer.

NITRATE LEVEL

1) Nitrate level can show changes in WATER POLLUTION.
2) SEWAGE and FERTILISERS can cause the nitrate level to RISE. This causes an increase in water pollution.

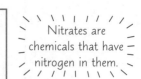

Nitrates are chemicals that have nitrogen in them.

CARBON DIOXIDE LEVEL

1) Carbon dioxide level can show changes in AIR POLLUTION.
2) An INCREASE in carbon dioxide could be caused by things like BURNING FOSSIL FUELS (see page 36).
3) More carbon dioxide means more air pollution.

...and Living Indicators

Living indicators are ORGANISMS that are affected by CHANGES in their ENVIRONMENT. For example:

MAYFLY NYMPHS

1) Mayfly nymphs are a type of insect.
2) They CAN'T LIVE where there's lots of WATER POLLUTION.
3) So if you find mayfly nymphs in a river, it shows that the WATER IS CLEAN.

Lichens

LICHENS

1) Lichens CAN'T LIVE where there's lots of AIR POLLUTION.
2) So if you find LOTS of lichens somewhere it shows that the AIR IS CLEAN.

PHYTOPLANKTON

1) Phytoplankton are tiny organisms that live in water.
2) Their numbers INCREASE when FERTILISERS and SEWAGE get into water.
3) So if you find LOTS of phytoplankton in water, it shows that the water has been POLLUTED.

Mmm, sewage fresh.

Practice Questions

1) Name one non-living indicator of pollution.
2) What are living indicators?
3) Name one living indicator of pollution. What type of pollution does it show?

Module B3 — Life on Earth

Sustainability

Sustainability's a tricky one to get your head around. Make sure you have a really good look at this page.

Sustainability is About Meeting People's Needs

People's needs are things like energy, water and food.

SUSTAINABILITY means letting people get what they need now — without harming the environment. This is so that people in future can still get what they need.

1) We're using up RESOURCES like fossil fuels. One day they will RUN OUT.
2) This means there won't be any left for people in the FUTURE to use.
3) So, using fossil fuels IS NOT SUSTAINABLE.

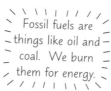

Fossil fuels are things like oil and coal. We burn them for energy.

1) SOME RESOURCES, like wind power, will NEVER RUN OUT.
2) This means people in the FUTURE will always be able to use them as well.
3) So, using wind power IS SUSTAINABLE.

Biodiversity Is An Important Part of Sustainability

LOSING BIODIVERSITY means that people in the future WON'T have the same resources that we do today. This means it's not sustainable.

EXAMPLE: MONOCULTURE CROP PRODUCTION

There's more on biodiversity on page 32.

① MONOCULTURE crop production is where farmers only grow ONE TYPE of crop in their fields. For example, only growing corn.

② It's BAD for biodiversity. This is because not many different species can survive if there's only one type of crop in a field.

③ This means it's NOT SUSTAINABLE.

I'm sure there used to be mice in this field...

Practice Questions

1) What does sustainability mean?
2) Is using fossil fuels sustainable?
3) "Monoculture crop production is good for biodiversity." True or false?

Module B3 — Life on Earth

Sustainability

There's just one more page to learn in this section — and it's really great. Honest...

Packaging can be Made More Sustainable

1) We usually throw away packaging like wrappers and boxes.
2) This ISN'T SUSTAINABLE.
3) We can make packaging MORE SUSTAINABLE by:

USING RENEWABLE MATERIALS

- A lot of packaging materials come from resources that will RUN OUT. For example, plastic is made from oil. This is NOT SUSTAINABLE.
- Using materials like paper and card (from trees) is MORE SUSTAINABLE.
- This is because these resources can be REPLACED once they've been used. For example, more trees can be planted.

CREATING LESS POLLUTION

- Most plastics AREN'T BIODEGRADABLE.
- This means they CAN'T be broken down naturally — so they pollute the land.
- Using BIODEGRADABLE materials (like cardboard) is MORE SUSTAINABLE. This is because they rot more easily.

USING LESS ENERGY

- Fossil fuels are burnt for ENERGY. Doing this DAMAGES the environment.
- Making packaging from RECYCLED materials uses LESS ENERGY than producing new materials.
- This means the environment ISN'T damaged as much because less energy is needed. This is MORE SUSTAINABLE.

4) The MOST sustainable thing to do is to USE LESS packaging material.
5) This is because:

Biodegradable materials still cause pollution. They take a while to break down.

Making and transporting any packaging material uses up energy.

Practice Questions

1) Give one reason why cardboard is more sustainable than plastic.
2) How can you use less energy when making packaging?
3) Give one reason why using less packaging is the most sustainable thing to do.

Module B3 — Life on Earth

Module C1 — Air Quality

How the Air was Made

The air between the ground and space is called the atmosphere. You need to know how it was made.

1. Volcanoes Gave Out Carbon Dioxide and Water Vapour

1) The Earth started out really HOT.

2) Volcanoes gave out CARBON DIOXIDE and WATER VAPOUR (steam).

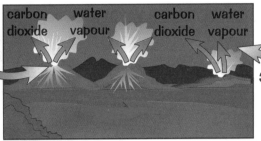

3) So the air was mostly MADE UP of carbon dioxide and water vapour.

2. The Water Vapour Cooled Down

1) Later on, the Earth COOLED DOWN.

2) The water vapour turned into LIQUID WATER. (This is called condensing.)

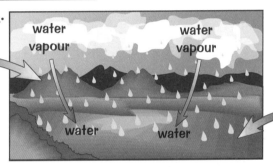

3) The liquid water made the OCEANS.

3. Carbon Dioxide Got Buried

Carbon dioxide was taken OUT of the air by PLANTS and OCEANS.

1) After a while, plants grew.
2) They took in CARBON DIOXIDE for photosynthesis.
3) They gave out OXYGEN by photosynthesis.
4) When the plants died some got BURIED.
5) After a long time they became FOSSIL FUELS.

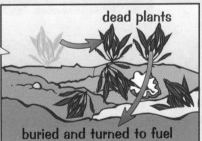

1) Lots of carbon dioxide dissolved in the OCEANS.

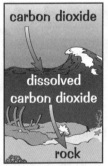

2) Some of it got buried in sedimentary ROCKS.

Sedimentary rocks are just a type of rock.

So the amount of carbon dioxide went DOWN and the amount of oxygen went UP.

Practice Questions

1) Where did the carbon dioxide and water vapour in the air come from?
2) What did the water vapour turn into?
3) What two things removed carbon dioxide from the air?

The Air Today

We take air for granted. But loads of things that we do every day are polluting and damaging it.

The Air is Made Up Of Different Gases

The air (atmosphere) surrounds the Earth. It contains:

1) Really tiny amounts of CARBON DIOXIDE, WATER VAPOUR, and some other gases.
2) A tiny bit of ARGON (1%)
3) Some OXYGEN (21%)
4) Mostly NITROGEN (78%)

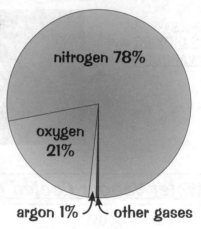

nitrogen 78%
oxygen 21%
argon 1% — other gases

Pollutant Gases Are Gases Added to the Air By...

1. Human Activity

Cars and power stations both give out pollutants:

2. Natural Processes

Volcanoes give out pollutants too

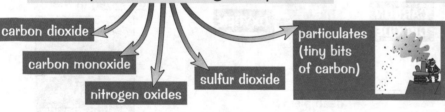

carbon dioxide
carbon monoxide
nitrogen oxides
sulfur dioxide
particulates (tiny bits of carbon)

Pollutants can be harmful. They:

1) HARM PEOPLE — for example, by causing carbon monoxide poisoning.
2) HARM the ENVIRONMENT — for example, by causing acid rain. This will harm people too.

Practice Questions

1) What gas makes up most of the atmosphere?
2) Name four pollutant gases.
3) Where do pollutant gases come from?

Module C1 — Air Quality

Chemical Reactions

When atoms swap around, that's a chemical reaction.

You Can Use Pictures to Show What Chemicals Are Made From

1) Everything is made from ATOMS. They are tiny. You can show them in pictures by drawing a blob.

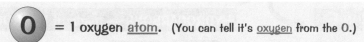 = 1 oxygen atom. (You can tell it's oxygen from the O.)

2) MOLECULES are atoms joined together.

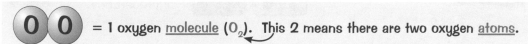 = 1 oxygen molecule (O_2). This 2 means there are two oxygen atoms.

3) Air is a MIXTURE of different gases. It can be drawn like this.
4) It's made of small MOLECULES.
5) The molecules have large SPACES between them.

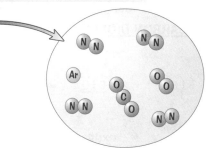

Chemical Reactions are When Atoms Swap Around

1) In a chemical reaction atoms are REARRANGED (swapped around) to make something new. For example:

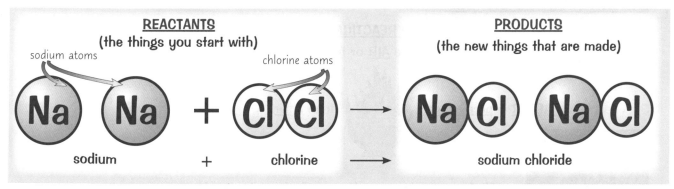

REACTANTS (the things you start with)

sodium atoms   chlorine atoms

Na Na + Cl Cl → Na Cl Na Cl

sodium + chlorine → sodium chloride

PRODUCTS (the new things that are made)

2) No atoms get LOST during the reaction, they just get REARRANGED.
3) For example, in the reaction above there are TWO chlorine atoms at the start and TWO at the end.
4) The products can have different PROPERTIES from the reactants — this means they can be very different.

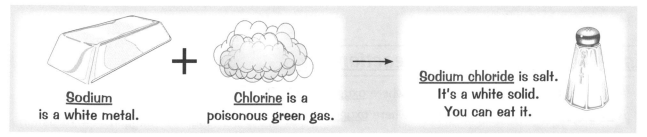

Sodium is a white metal.   Chlorine is a poisonous green gas.   Sodium chloride is salt. It's a white solid. You can eat it.

Practice Questions

1) What are molecules?
2) Circle the right word to complete this sentence:
   air is made of small molecules with SMALL / L A R G E spaces in between them.
3) A chemical reaction starts with 54 321 atoms. How many are there at the end of the reaction?

Module C1 — Air Quality

Fuels

Fuels are important. Luckily, here's a page all about them. I can tell by the word "Fuels" above.

Burning Fuels is A Chemical Reaction

1) There are different types of FUEL.

COAL — made of CARBON C	HYDROCARBONS — made of HYDROGEN H and CARBON C
 coal	petrol diesel Petrol, diesel and fuel oil are all hydrocarbons.  fuel oil

2) When a fuel burns its atoms join with OXYGEN.
3) CARBON DIOXIDE is made. Sometimes WATER is made too. For example:

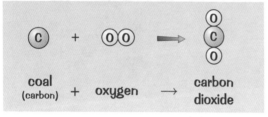

coal (carbon) + oxygen → carbon dioxide

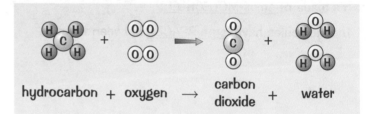

hydrocarbon + oxygen → carbon dioxide + water

Burning Needs Oxygen

1) Burning is also called a COMBUSTION REACTION.
2) Oxygen for burning can come from the AIR or from PURE OXYGEN.

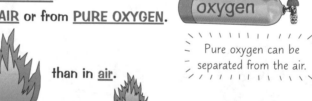

Pure oxygen can be separated from the air.

3) Fuels burn FASTER in pure oxygen than in air.

4) Pure oxygen is used in oxy-fuel welding torches.

Know These Two Sorts of Reaction

1) An OXIDATION reaction is a reaction where oxygen is ADDED.
2) A REDUCTION reaction is a reaction where oxygen is LOST.

fuel + oxygen = oxidation

Practice Questions

1) What is a hydrocarbon made of?
2) What chemicals are made when a hydrocarbon burns?
3) Is adding oxygen to atoms a reduction reaction?

Module C1 — Air Quality

Air Pollution — Carbon

Coal, petrol, diesel and oil are all types of fossil fuel. This page is about the pollution they cause.

Carbon Dioxide, Carbon Monoxide and Carbon Particles Are Pollution

Cars and power stations burn fossil fuels.
This pollutes the atmosphere with:
1) CARBON DIOXIDE (CO_2)
2) CARBON MONOXIDE (CO)
3) PARTICULATE CARBON (C)

This just means tiny bits of carbon.

The pollution STAYS in the air unless it gets removed.

Particulate carbon and carbon particles are the same thing.

Carbon dioxide DISSOLVES in rainwater and in seas.

PLANTS take in carbon dioxide from the air when they photosynthesise.

Carbon particles fall down as SOOT. This makes buildings look dirty.

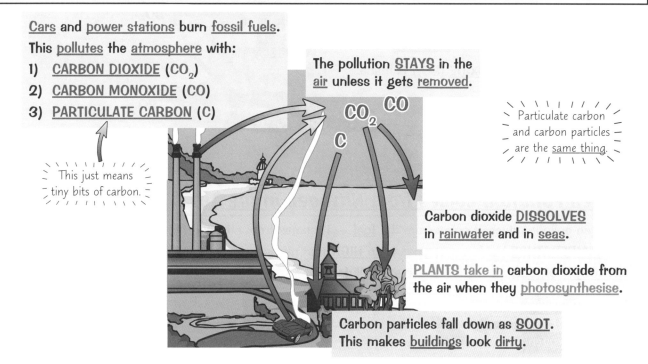

Carbon Monoxide and Carbon are Made by Incomplete Burning

1) If there's NOT MUCH oxygen when fuel is burnt then you get something called INCOMPLETE BURNING.
2) This is how the carbon monoxide and carbon particles are made:

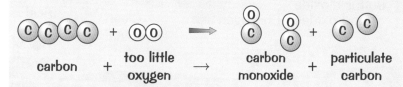

carbon + too little oxygen → carbon monoxide + particulate carbon

Make Sure You Know These Molecules

NAME	DIAGRAM	FORMULA
water	(O)(H)(H)	H_2O
carbon dioxide	(O)(C)(O)	CO_2
carbon monoxide	(C)(O)	CO

The LETTERS in a formula show what SORTS of atoms there are. The NUMBERS show HOW MANY of each sort there are.

CO_2 — 1 carbon atom, 2 oxygen atoms

Practice Questions

1) What is the name of this molecule? What is its formula?
2) What is soot made up of?
3) How is carbon dioxide removed from the air?

Module C1 — Air Quality

Air Pollution — Sulfur and Nitrogen

Cars and power stations pollute the air with sulfur and nitrogen.

Sulfur Pollution Comes from Fuels

1) Fossil fuels sometimes have SULFUR (S) in them.
2) When a fuel BURNS the sulfur burns too.
3) The pollutant SULFUR DIOXIDE (SO_2) is made. It goes into the air.

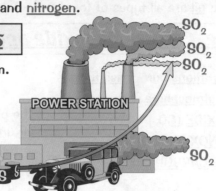

Nitrogen Pollution Comes from Nitrogen in the Air

1) Nitrogen pollution doesn't come from the fuel — it comes from nitrogen (N) in the AIR.
2) The TEMPERATURE inside car engines is so HIGH that nitrogen and oxygen REACT.
3) NITROGEN OXIDES are made. These are pollutants.

Sulfur and Nitrogen Pollution Causes Acid Rain

The gases react with water and oxygen to make ACID RAIN.

Acid rain kills plants and animals.

Make Sure You Know These Molecules

NAME	DIAGRAM	FORMULA
sulfur dioxide	S, O, O	SO_2
nitrogen monoxide	N, O	NO
nitrogen dioxide	N, O, O	NO_2

These are both types of NITROGEN OXIDE.

Practice Questions

1) Where does the nitrogen in nitrogen pollution come from?
2) What problem does sulfur and nitrogen pollution cause?

Module C1 — Air Quality

Reducing Pollution

There are lots of things we can do to try to reduce air pollution.

We Can Reduce Pollution From Power Stations

We can do this in THREE ways:

①
- Using less ELECTRICITY.
- This means less fossil fuels are burned.
- So less CARBON DIOXIDE is given out.

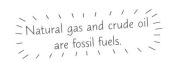
Natural gas and crude oil are fossil fuels.

②
- Taking sulfur out of the natural gas and fuel oil that power stations use.
- This means less SULFUR DIOXIDE is given out.

Sulfur-free

③
- Cleaning the gas coming out of coal-burning power stations. (These gases are called FLUE GASES.)
- This means less SULFUR DIOXIDE and PARTICULATE CARBON is given out.

We Can Reduce Pollution From Our Cars

We can do this in FIVE ways:

①
- Using more EFFICIENT engines.
- They burn less fuel and make less pollution.

 OLD  EFFICIENT

② CARS — PUBLIC TRANSPORT

- Using PUBLIC TRANSPORT.
- This burns less fuel and makes less pollution.

③
- Using LOW-SULFUR fuel.
- This means less sulfur dioxide is given out.

Low Sulfur

④
- Putting CATALYTIC CONVERTERS in cars.
- They remove carbon monoxide and nitrogen monoxide from the gases the car gives out (exhaust gases).

CARBON MONOXIDE in → out CARBON DIOXIDE
NITROGEN MONOXIDE in → out NITROGEN

⑤
- Having LAWS on car emissions (pollution from a car's exhaust).
- Cars that make too much pollution will fail their MOT tests.

Practice Questions

1) What do catalytic converters do?
2) What happens to cars that make too much pollution?
3) How does using less electricity help to reduce pollution?

Module C1 — Air Quality

Module C2 — Material Choices

Natural and Synthetic Materials

This section is all about materials and what they can be used for.

All Materials are Made Up of Chemicals

1) We use loads of different materials, like:

   METALS   POLYMERS   CERAMICS

2) Every material is made up of chemicals. Chemicals are lots of atoms joined together.

3) Some materials are MIXTURES of chemicals. For example, rock salt is made up of salt and sand.

salt + sand = rock salt

Some Materials are Natural

A lot of the materials that we use are made by LIVING THINGS. For example:

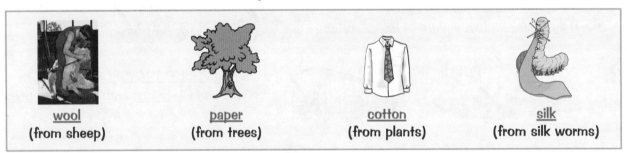

wool (from sheep)   paper (from trees)   cotton (from plants)   silk (from silk worms)

Some Materials are Made by Humans

Materials made by HUMANS are called SYNTHETIC MATERIALS.

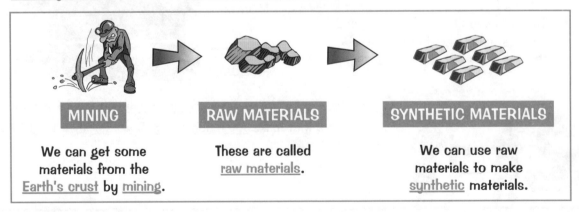

MINING — We can get some materials from the Earth's crust by mining.

RAW MATERIALS — These are called raw materials.

SYNTHETIC MATERIALS — We can use raw materials to make synthetic materials.

Practice Questions

1) What are materials made of?
2) Name three natural materials.

Materials and Properties

Not all materials are the <u>same</u>. A hat made out of <u>spaghetti hoops</u> won't be as good as one made of <u>plastic</u>.

Different Materials Have Different Properties

A material's <u>properties</u> are what it's <u>like</u>. Here are the properties you need to <u>know about</u>.

MELTING POINT

This is the <u>temperature</u> where a <u>SOLID</u> turns to a <u>LIQUID</u>.

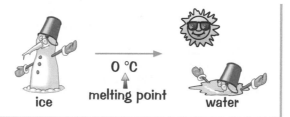

STRENGTH

1) You can see how <u>strong</u> something is by seeing how easy it is to <u>BREAK IT</u> or <u>CHANGE ITS SHAPE</u>.
2) There are <u>two</u> types of strength you need to know about:

<u>TENSION STRENGTH</u> — how much a material can stand up to a <u>pulling force</u>.

Pull

<u>COMPRESSIVE STRENGTH</u>: how much a material can stand up to a <u>pushing force</u>.

Push

STIFFNESS

1) A <u>stiff</u> material is good at <u>NOT bending</u>.

2) A <u>bendy</u> material can <u>still be strong</u> if it goes back to its <u>starting shape</u> when you <u>stop</u> bending it.

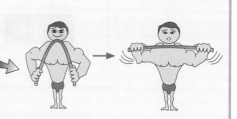

HARDNESS

1) The hardness of a material is how <u>difficult</u> it is to <u>CUT</u>.
2) The <u>hardest</u> material in <u>nature</u> is <u>diamond</u>.

DENSITY

Density is <u>how much STUFF</u> there is in a certain <u>amount of space</u>.

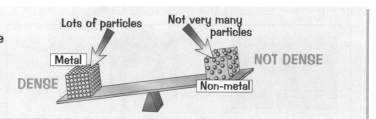

Practice Questions

1) Name the <u>two</u> types of <u>strength</u>.
2) What is the <u>hardness</u> of a material?
3) What is <u>density</u>?

Module C2 — Material Choices

Materials, Properties and Uses

Materials all have different <u>properties</u> that describe what the material is like.

The Uses of a Material Depend on What It's Like

1) Every material has different PROPERTIES.
2) <u>Properties</u> are <u>what a material is LIKE</u>. For example, metals are <u>shiny</u>.
3) You need to be able to look at the <u>properties</u> of a material and work out what it could be <u>USED</u> for.

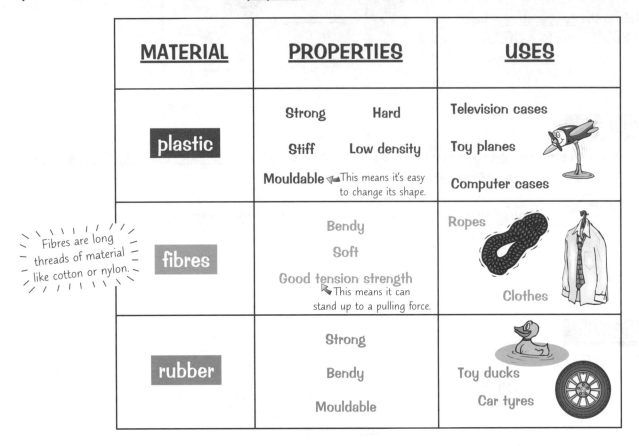

MATERIAL	PROPERTIES	USES
plastic	Strong Hard Stiff Low density Mouldable ← This means it's easy to change its shape.	Television cases Toy planes Computer cases
fibres	Bendy Soft Good tension strength ↖ This means it can stand up to a pulling force.	Ropes Clothes
rubber	Strong Bendy Mouldable	Toy ducks Car tyres

Fibres are long threads of material like cotton or nylon.

How Good a Product is Depends on What It's Made From

1) All <u>products</u> are made to do a certain <u>job</u>.
2) The <u>EFFECTIVENESS</u> of a product is <u>how good it is</u> at this job.
3) How good a product is depends on the <u>materials</u> it's made from.
4) The <u>materials</u> it's made from also affects how <u>LONG</u> a product <u>will LAST</u> (its <u>DURABILITY</u>).

A product is something that is made for us to use — like a bath or a car.

 vs  Metal sculptures will last much longer than ice sculptures.

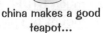
china makes a good teapot...   ... chocolate doesn't.

Practice Questions

1) What is the <u>effectiveness</u> of a product?
2) Plastics are <u>hard</u>, <u>strong</u> and <u>stiff</u>. Rubber is <u>soft</u> and <u>bendy</u>.
   Would you make a <u>tyre</u> out of plastic or rubber? Explain your answer.

Module C2 — Material Choices

51

Crude Oil

Crude oil is made from plants and animals that were buried thousands of years ago.

Crude Oil is a Mixture of Hydrocarbons

1) Crude oil is made up of a MIXTURE of HYDROCARBONS.
2) Hydrocarbons are chains made of CARBON and HYDROGEN atoms only.

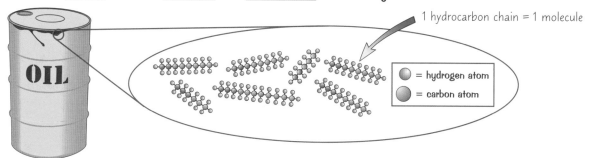

1 hydrocarbon chain = 1 molecule

○ = hydrogen atom
● = carbon atom

The Chains Need Energy to Break Apart

You don't always have to show the hydrogens when you draw hydrocarbons.

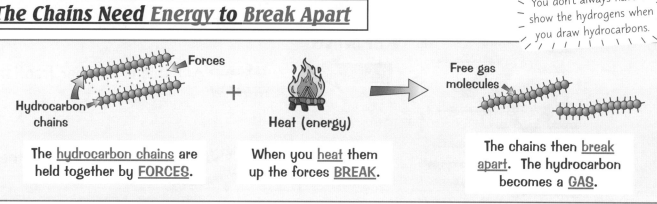

Hydrocarbon chains — Forces + Heat (energy) → Free gas molecules

The hydrocarbon chains are held together by FORCES.

When you heat them up the forces BREAK.

The chains then break apart. The hydrocarbon becomes a GAS.

SHORT HYDROCARBONS

1) Short hydrocarbons DON'T have many forces to hold them together.
2) So NOT MUCH energy is needed to turn them from a liquid to a gas.
3) This means they have LOW BOILING POINTS.

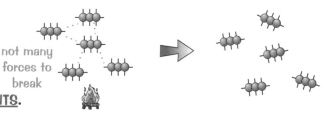

not many forces to break

LONG HYDROCARBONS

1) Long hydrocarbons have LOTS of forces to keep the molecules together.
2) So it takes LOADS of energy to break them apart.
3) This means they have HIGH BOILING POINTS.

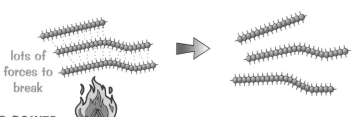

lots of forces to break

Practice Questions

1) Name the two types of atom hydrocarbons are made of.
2) How are the hydrocarbon chains held together?
3) Why do long hydrocarbons have high boiling points?

Module C2 — Material Choices

Uses of Crude Oil

People will pay loads of money for crude oil because it can be used to make loads of stuff.

Crude Oil is Separated into Fractions

1) The hydrocarbons in crude oil can be SPLIT UP into GROUPS.
2) The hydrocarbons in a group have about the SAME BOILING POINT.
3) Each group is called a FRACTION.
4) Separating crude oil like this is called REFINING.
5) It's done by the petrochemical industry using FRACTIONAL DISTILLATION.

Hydrocarbons in different fractions are different lengths.

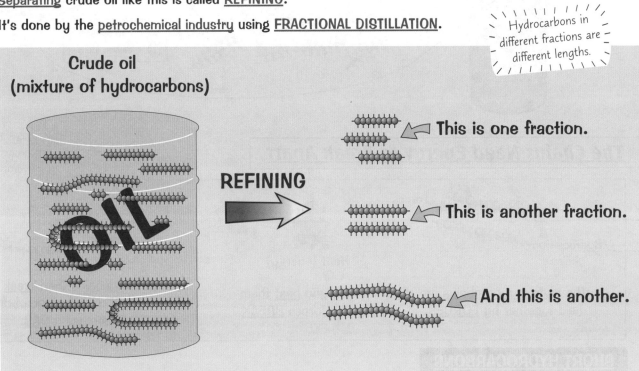

Crude Oil Has Lots of Uses

The different fractions are used for different things. They can be used as:

RAW MATERIALS: These are used to make new chemicals. Making new chemicals is called chemical synthesis.

LUBRICANTS: These are used to make machinery run smoothly.

FUELS: Most crude oil is used for fuels.

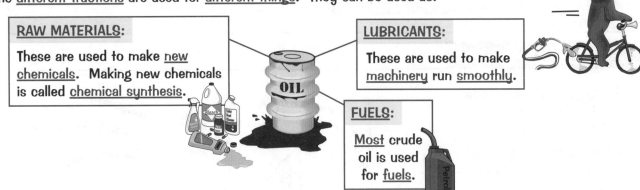

Practice Questions

1) What is separating crude oil called?
2) Is the following sentence true or false? The hydrocarbons in a group are about the same length.
3) Give two uses of crude oil fractions.

Module C2 — Material Choices

Polymerisation

Polymers are very important man-made materials. You can make loads of useful stuff from them.

Polymers are Made From Loads of Small Molecules

1) Some small molecules are called MONOMERS.
2) These can join together to make very long molecules called POLYMERS.
3) This reaction is called POLYMERISATION.

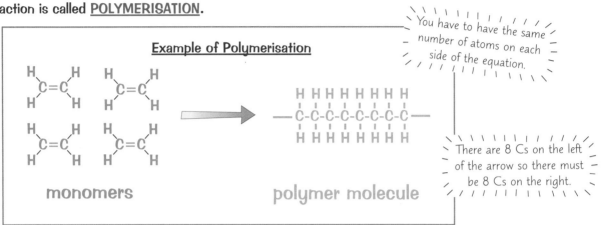

You have to have the same number of atoms on each side of the equation.

There are 8 Cs on the left of the arrow so there must be 8 Cs on the right.

There are Lots of Different Types of Polymers

1) Different polymers can have very DIFFERENT PROPERTIES.
2) Using different polymers you can make loads of different materials.

Some are strong and stiff.

Some are light and stretchy.

Polymers can be Used Instead of Older Materials

1) Some polymers have better properties than older materials.
2) So we now use them instead of the older materials. For example:

OLDER MATERIAL	POLYMER	WHY IS THE POLYMER BETTER?
Cotton, wool or silk	Nylon	It's lighter, cheaper and lasts much longer.
Wood	PVC	It's strong, it lasts longer and it's better in extreme weather.

Practice Questions

1) What are polymers made from?
2) Do all polymers have the same properties?
3) Give an example of a polymer that is used now instead of a more traditional material.

Module C2 — Material Choices

Structure and Properties of Polymers

What a polymer is like depends on the way it's made.

Polymers are Made From Lots and Lots of Polymer Chains

Polymer chains are held together by FORCES.

Some chains are held together by WEAK FORCES.

1) Only a bit of energy is needed to separate the chains.

2) So the polymer will have a LOW melting point.
3) It will also be EASY TO STRETCH.

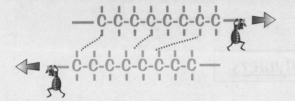

Some chains are held together by STRONG FORCES.

1) Lots of energy is needed to separate the chains.

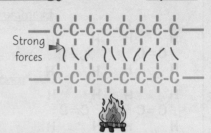

2) These polymers have HIGHER melting points.
3) They CAN'T BE EASILY STRETCHED.

Polymers Can be Changed to Give Them Different Properties

CHANGE	CHANGE IN PROPERTIES
LONGER CHAIN LENGTH —c-c-c-c-c— ➡ —c-c-c-c-c-c-c-c-c-c-c— Short chain Long chain	• Polymers with LONGER chains are STIFFER. • They also have HIGHER melting points.
CROSS-LINKING polymer chain cross-linking	• You can add chemicals to hold the chains together. • This makes the polymer STIFFER and STRONGER.

Practice Questions

1) A polymer is held together with strong forces. Will it have a high or low melting point?
2) A polymer is held together with weak forces. Will it be easy to stretch?
3) Brendan makes a polymer which is soft and flexible. How could Brendan make his polymer stronger?

Module C2 — Material Choices

Nanotechnology

Just time to squeeze in something really small before the end of the section...

Nanomaterials Are Really Really Really Really Tiny
...smaller than that.

1) Really tiny particles are called 'NANOPARTICLES'. *You could fit over 100 000 nanoparticles on this full stop.*
2) They are about 1-100 nanometres across.

Nanoparticles are Made in Different Ways

HOW NANOPARTICLES ARE MADE	EXAMPLE
BY NATURE	• SEASPRAY — the sea makes salt nanoparticles.
BY ACCIDENT	• COMBUSTION — when fuels are burnt, soot nanoparticles are made.
BY SCIENTISTS	• Nanoparticles can also be made in labs by SCIENTISTS.

Nanoparticles can be Added to Other Materials

Nanoparticles can be ADDED to materials to give them DIFFERENT PROPERTIES. For example:

- Nanoparticles are added to PLASTICS used in sports, e.g. tennis rackets.
- They make the plastic much STRONGER.

- Silver nanoparticles are added to SURGICAL MASKS and bandages.
- They help to KILL BACTERIA.

Nanoparticles Might be Harmful

1) We still don't know how nanoparticles affect the BODY.
2) So we have to TEST anything with nanoparticles in to check they won't cause any HARM.
3) Some people are worried that products with nanoparticles in haven't been fully tested.

Practice Questions

1) Is the following sentence true or false? Nanoparticles are 1-100 nanometres across.
2) Name a nanoparticle which is made naturally.
3) Scientists have added nanoparticles to surgical masks. How does this make the mask better?

Module C2 — Material Choices

Tectonic Plates

We can tell a lot about what the Earth was like thousands of years ago by looking at it today...

The Earth's Surface is Made Up of Tectonic Plates

1) The Earth is made up of different LAYERS.
2) On the outside is the CRUST.

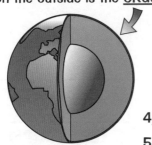

3) The crust is broken up into large pieces called TECTONIC PLATES.

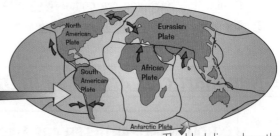

The black lines show the edges of the tectonic plates.

4) These plates move around very very SLOWLY.
5) The crust that makes up Britain has moved over the surface of the Earth.

Some Rocks are Magnetic

Scientists that study rocks are called geologists.

New crust made

1) At some places NEW CRUST is pushed up between tectonic plates.
2) This new crust is MAGNETISED.
3) Scientists look at the rock to work out how the tectonic plates have MOVED. The magnetic rock gives them CLUES.

Rocks can Tell Us About the Earth's History

1) The rocks in Britain were made in different CLIMATES.
2) Scientists look at SEDIMENTARY ROCKS (a type of rock) to learn about what it was like where they were made.

cold climate   hot climate

SCIENTISTS LOOK FOR FOSSILS, SHELLS AND RIPPLES IN ROCKS

- FOSSILS are what's left of dead plants and animals.
- Fossils can tell you what it was like when the rock was made.

This fossil of a footprint tells you that the rock was formed on land.

This fossil of a shell tells you that the rock was made underwater.

- Rocks made UNDERWATER can contain shells.
- They can also have ripples on them made by the sea or rivers.

Practice Questions

1) The Earth's crust is broken up into large pieces. What are these pieces called?
2) Why do scientists look at magnetic rock?
3) A rock has a fossil of a fish in it and has ripples on it. Was it made on land or underwater?

Resources in the Earth's Crust

There's loads of money to be made from the stuff we find in the Earth's crust — like coal, oil and salt.

Coal, Limestone and Salt are Found in the Earth's Crust

1) Natural materials we can use are called RESOURCES.
2) There are different types. For example:
3) Some resources can be used to make CHEMICALS.
4) Factories for making chemicals are usually built NEAR to the resources.
5) There's a CHEMICAL INDUSTRY in the north-west of England because there's LIMESTONE, SALT and COAL nearby.

The chemical industry is all the businesses that make chemicals.

Limestone is Made From Sediment

1) SEDIMENT is SMALL PARTICLES of natural materials.
2) Sediment can be made from shells and bones of SEA CREATURES.
3) It's also made when the wind wears away rocks. This is called EROSION.
4) Layers of sediment get BURIED under more layers.
5) The weight pressing down turns the layers at the bottom into ROCK.

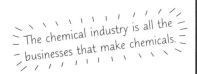

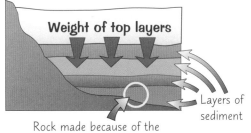

Rock made because of the weight of the layers above.

Coal Can be Made When the Pressure and Temperature are High

1) Some coal is made when the PRESSURE and TEMPERATURE are high.
2) This can happen when MOUNTAINS are made (mountain building).

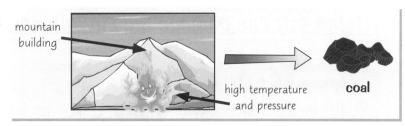

Salt is Found in Sea Water

1) SEAS are made from salt water.
2) All the water EVAPORATED from some old seas.
3) The salt was left behind in the Earth's CRUST.

Evaporated means turned from a liquid to a gas.

Practice Questions

1) Name three resources.
2) What is limestone made from?
3) What resource is made when sea water evaporates?

Module C3 — Chemicals in Our Lives

Salt

You can get salt out of the ground in two different ways. I don't mind — as long as I can have it on my chips.

Salt is Found Underground

1) Build-ups of salt are called SALT DEPOSITS.
2) There are lots of salt deposits UNDERGROUND.
3) You can use NORMAL MINING or SOLUTION MINING to get the salt out.

Normal Salt Mining — Digging the Salt Out

1) The salt is drilled, blasted and dug out.
2) Most salt mined this way is used to GRIT ROADS.
3) The salt is also used in FOOD or for MAKING CHEMICALS.

Solution Mining is Another Way of Salt Mining

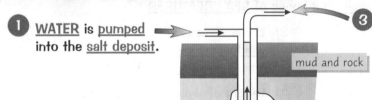

1) WATER is pumped into the salt deposit.
2) The water DISSOLVES the salt.
3) Pipes bring the salt water up to the SURFACE.

4) Most TABLE SALT and the salt used for MAKING CHEMICALS is made this way.

Mining Salt Can Damage the Environment

1) Land above old mines can collapse into the HOLES.
2) This can DAMAGE buildings near the mines so people have to PAY to fix them.

1) Mining needs a lot of ENERGY.
2) This comes from burning fossil fuels.
3) This uses up RESOURCES and causes POLLUTION.

You Can Also Get Salt From the Sea

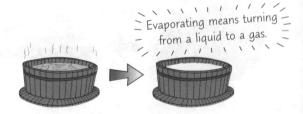

1) In HOT countries you can get salt by evaporating seawater.
2) Seawater is put in tanks and left in the SUN.
3) The water evaporates. The SALT is left behind.

Evaporating means turning from a liquid to a gas.

Practice Questions

1) You can dig salt out of the ground. Give one more way that we can get salt out of the ground.
2) Give two ways that mining can affect the environment.
3) How can you get salt from the sea?

Module C3 — Chemicals in Our Lives

Salt in the Food Industry

Salt is jolly useful stuff — especially when it's added to food.

Salt *is* Added to Lots of Foods

There are lots of ADVANTAGES and DISADVANTAGES to adding salt to food.

ADVANTAGES	DISADVANTAGES
TASTIER FOOD • Salt is used as a FLAVOURING. • This means it's added to food to make it TASTE BETTER.	**RISK OF HEALTH PROBLEMS** • Eating too much salt may cause HIGH BLOOD PRESSURE. • High blood pressure can lead to STROKES and HEART ATTACKS. • Eating too much salt could also increase the chance of getting stomach cancer, osteoporosis (weak bones) and kidney failure.
FOOD LASTS LONGER • Salt is used as a PRESERVATIVE. • This means it will make food LAST LONGER.	

The Government Gives Out Information On Food

1) There are two main government departments that give advice on food safety...

- The Department of Health
- The Department for Environment, Food and Rural Affairs

2) As part of their job they:

 Tell the public about how food affects their HEALTH.

- Look at the health risk of chemicals in food to make sure they're SAFE.
- This is called a RISK ASSESSMENT.

Labels Give You Information On Food

The LABELS on food tell you HOW MUCH SALT is in them.

NUTRITIONAL INFORMATION	per packet	per 100 g
FAT	11.0 g	15.4 g
PROTEIN	2.1 g	6.2 g
SALT	0.5 g	1.6 g

- In this packet of crisps there is 0.5 g of SALT.
- There is 1.6 g of SALT in every 100 g of crisps.

Practice Questions

1) Give two advantages of adding salt to food.
2) Give one disadvantage of adding salt to food.
3) Name one government department that gives advice on food safety.

Module C3 — Chemicals in Our Lives

Electrolysis of Salt Solution

OK, I get it, salt is really useful. Don't worry though, this is the last page on the stuff...

Salt *is* Used to Make Chemicals

1) <u>Salt</u> is important for <u>making CHEMICALS</u>.
2) This means it's <u>very important</u> for the <u>CHEMICAL INDUSTRY</u>.
3) You can get useful chemicals <u>out of salt</u> using <u>ELECTROLYSIS</u>.
4) In electrolysis you <u>pass an electric current</u> through <u>salt solution</u>.
5) The salt solution splits into <u>CHLORINE</u>, <u>HYDROGEN</u> and <u>SODIUM HYDROXIDE</u>.

Salt solution and brine are just fancy words for salty water.

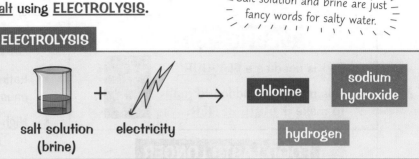

The Chemicals Made by Electrolysis are Really Useful

CHEMICAL	USES
CHLORINE	1) disinfectants 2) killing bacteria 3) household bleach
HYDROGEN	1) making the chemical ammonia 2) margarine 3) welding and metal cutting
SODIUM HYDROXIDE	1) soap 2) oven cleaner 3) household bleach

Electrolysis of Salt Water can Harm the Environment

1) Electrolysis needs a <u>lot of ENERGY</u>.
2) This energy comes from burning <u>fossil fuels</u>.
3) Burning fossil fuels puts <u>POLLUTANTS</u> like <u>carbon dioxide</u> into the air — this can <u>damage the ENVIRONMENT</u>.

Practice Questions

1) What is another name for <u>salt solution</u>?
2) Name <u>three</u> chemicals you can make by <u>electrolysis of salt solution</u>.
3) Give <u>one</u> way that electrolysis of salt solution can <u>harm the environment</u>.

Module C3 — Chemicals in Our Lives

Chlorination

It's easy to take clean water for granted... turn on the tap, and there it is. It's all down to chlorination.

Chlorine is Used to Make Water Safe

1) In the UK, CHLORINE is added to drinking water to make it SAFE.
2) Chlorine is used because:

 It stops algae growing. It kills bacteria. It gets rid of bad tastes and smells.

3) Adding chlorine to water is called CHLORINATION.

Chlorine Can be Made from Hydrogen Chloride

1) Chlorine is found in COMPOUNDS.
2) You can get chlorine out of hydrogen chloride by reacting it with OXYGEN.

Hydrogen chloride is a compound (it's got two different elements in it).

HYDROGEN CHLORIDE
↑ ↑
hydrogen and chlorine

 + OXYGEN ⟶

Colourless gas with white fumes.
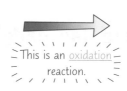 This is an oxidation reaction.
Green, smelly gas. Will kill bacteria.

3) You have to take the chlorine OUT of the hydrogen chloride before it can be used.
4) This is because CHLORINE will KILL bacteria, but COMPOUNDS with chlorine IN might NOT.

Chlorination is Great for Your Health

1) Before chlorination, lots of people became very ILL from drinking DIRTY WATER.
2) Since chlorination was started public health has got a lot better.
3) This is because chlorination means there's CLEAN WATER for everyone to drink.

4) There are also some disadvantages to chlorination.
5) Some chemicals in water can react with chlorine to make harmful chemicals.
6) But, the chance of this happening is really SMALL.

Practice Questions

1) Give two reasons why chlorine is added to drinking water.
2) Give one disadvantage of using chlorination.

Module C3 — Chemicals in Our Lives

Alkalis

Alkalis are really important chemicals — you'll find them at home, in the lab and in industry.

Alkalis are a Type of Chemical

1) When an ACID and an ALKALI react together they make a SALT and WATER.

$$\text{Acid} + \text{Alkali} \rightarrow \text{Salt} + \text{Water}$$

2) This is called a NEUTRALISATION REACTION.

The salt we eat is only one type of salt. In fact there are loads of different types.

Alkalis have been Used for Hundreds of Years

Alkalis are used for all sorts of things:

Neutral is when something's not an acid or an alkali.

Dyes are what give clothes their colour.

Alkalis help DYES stick to cloth.

Alkalis can help turn fats and oils into SOAP.

Alkalis are used to make GLASS.

Farmers use alkalis to make acidic soils NEUTRAL.

Today Alkalis are Man-Made

The alkalis we use TODAY are very different from the ones used in the PAST.

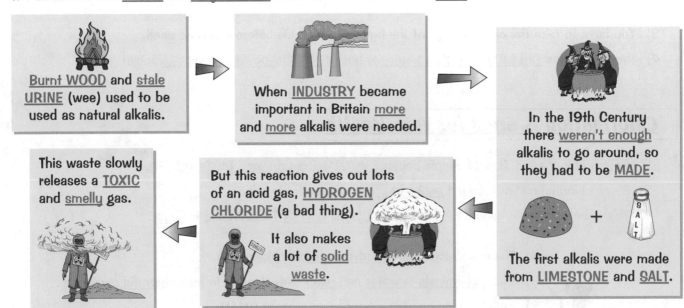

Burnt WOOD and stale URINE (wee) used to be used as natural alkalis.

This waste slowly releases a TOXIC and smelly gas.

When INDUSTRY became important in Britain more and more alkalis were needed.

But this reaction gives out lots of an acid gas, HYDROGEN CHLORIDE (a bad thing). It also makes a lot of solid waste.

In the 19th Century there weren't enough alkalis to go around, so they had to be MADE.

The first alkalis were made from LIMESTONE and SALT.

Practice Questions

1) What two things do you get when you react an acid with an alkali?
2) Give two uses of alkalis.
3) Why did we have to start making alkalis?

Module C3 — Chemicals in Our Lives

Impacts of Chemical Production

Time to look at making chemicals — including the times when things go wrong...

Lots of Things Can be Made Using Chemistry

1) Chemicals can be used to make lots of different things...

2) As there are so many chemicals, they can't all be TESTED as much as we'd like.
3) This means we don't know if some chemicals could harm the ENVIRONMENT or PEOPLE'S HEALTH.

Some Chemicals Stay in the Environment for a Long Time

1 Some chemicals end up in LAKES or RIVERS or are eaten by ANIMALS.

2 These chemicals could be carried LARGE DISTANCES as the water and animals move around.

3 They could be passed along the FOOD CHAIN and cause harm to other animals and even HUMANS.

Plasticisers can Harm the Environment

1) PVC is a POLYMER (see pages 53-54).
2) It's made of CARBON, HYDROGEN and CHLORINE.
3) Chemicals called PLASTICISERS are added to PVC.
4) Plasticisers can LEAK OUT of the PVC and into water nearby.

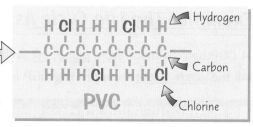

5) These plasticisers can be HARMFUL.
6) They can build up in animals like fish and end up being eaten by HUMANS.

Practice Questions

1) Why can't we test all chemicals as much as we'd like to?
2) What three elements is PVC made of?
3) Explain how chemicals poured into a river could get into our food.

Module C3 — Chemicals in Our Lives

Life Cycle Assessments

If a company wants to make a new product, they carry out a life cycle assessment (LCA).

Life Cycle Assessments Have Four Stages

Life cycle assessments look at FOUR different stages in the LIFE of a product.

Life cycle assessment can be written as LCA.

Life Cycle Assessments Show Environmental Effects

A life cycle assessment looks at THREE things for each stage of the product's life. It looks at:

THE USE OF RESOURCES
This is what materials are used. For example, water, metals or oil.

THE ENERGY USED OR MADE
This is what energy is used or given out. For example, electricity or fossil fuels.

HOW THE ENVIRONMENT IS AFFECTED
This is whether the environment will be damaged. For example, by waste and pollution.

Example: The Life Cycle Assessment of a CGP Book

A CGP book is made from paper, ink and love. The book is made at the CGP factory.
All the waste paper from printing CGP books is recycled.

The MAKING THE PRODUCT stage of a CGP book life cycle assessment.

What resources are used? Paper and ink are used up.

Is energy is used or made? 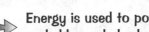 Energy is used to power the factory. The energy was probably made by burning fossil fuels — this causes pollution.

Is the environment damaged? 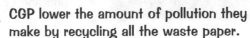 CGP lower the amount of pollution they make by recycling all the waste paper.

A full life cycle assessment would have to look at all the other stages too.

Practice Questions
1) What are the four stages of the life cycle assessment of a product?
2) What are the three things you have to think about when you're doing a life cycle assessment?

Module C3 — Chemicals in Our Lives

Module P1 — The Earth in the Universe

The Solar System

The Earth is just one part of the Solar System.

These Things Are All in the Solar System

The SOLAR SYSTEM has all these things in it:

- The SUN is in the middle.
- PLANETS and DWARF planets ORBIT (go round) the Sun.
- Dwarf means smaller than normal.
- Eight planets go round the Sun. The EARTH is the third one along.
- COMETS are small bits of rock and ice that go round the Sun.
- MOONS go round planets.
- ASTEROIDS are small bits of rock that go round the Sun.

The Solar System is 5 Thousand Million Years Old

1) The Solar System was made 5000 million years ago.
2) It was made over a LONG TIME from big clouds of DUST and GAS.

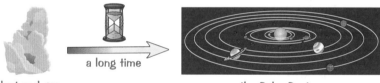

dust and gas — a long time — the Solar System

The Sun's Energy comes from Fusion

1) The Sun and other STARS are made of hydrogen and helium gas.
2) Hydrogen atoms FUSE (join) together to make helium inside stars.
3) This is where a star's HEAT and LIGHT energy comes from.

The Sun is a star.

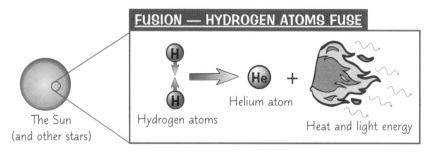

FUSION — HYDROGEN ATOMS FUSE

The Sun (and other stars) — Hydrogen atoms → Helium atom + Heat and light energy

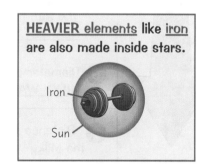

HEAVIER elements like iron are also made inside stars.

Iron — Sun

Practice Questions

1) a) What do planets go round?
   b) What do moons go round?
2) What fuses together to make helium inside stars?

Beyond the Solar System

The Universe is everything there is and it's HUGE.

We're in the Milky Way Galaxy

1) GALAXIES are big collections of stars.
2) Our Sun is one of thousands of millions of STARS in the Milky Way GALAXY.
3) There are thousands of millions of galaxies in the UNIVERSE.

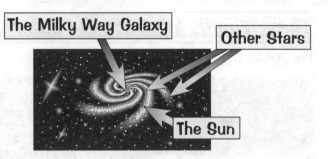

We Use Light Years for Distances in Space

1) Light travels really fast — 300 000 km/s in space.
2) We use LIGHT YEARS to measure distances in space:

ONE LIGHT YEAR IS THE DISTANCE LIGHT TRAVELS IN ONE YEAR.

You Need to Know These Sizes, Distances and Ages

You need to be able to put the sizes of these things in the Universe in order:

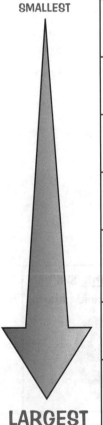

SMALLEST → LARGEST

1. Diameter of the EARTH
2. Diameter of the SUN
3. Diameter of the Earth's ORBIT
4. Diameter of the SOLAR SYSTEM
5. Distance from the Sun to the NEAREST STAR
6. Diameter of the MILKY WAY GALAXY
7. Distance from the Milky Way to the NEAREST GALAXY

Diameter is the distance from one side to the other.

You need to know some ages too:

Different stuff in space	Age
Earth	5000 million years
Sun	5000 million years
Universe	14 000 million years

Practice Questions

1) What galaxy is our Sun in?
2) Which is larger, the diameter of the Milky Way or the diameter of the Solar System?

Module P1 — The Earth in the Universe

Looking Into Space

We can't travel TO stars to study them. So we study the light coming from them.

Light Tells Us A Lot About Stars and Galaxies

1) Stars give out LIGHT and other electromagnetic radiation.
2) How MUCH light a star gives out is called its brightness.
3) How bright a star looks from Earth depends on TWO things:
   - how bright it really is,
   - how far away it is from Earth.

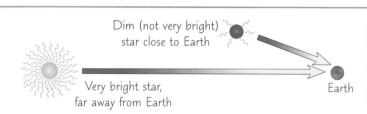

The DIM star is CLOSER than the BRIGHT star. But both stars look as bright as each other from Earth.

4) You can tell how FAR AWAY a star is by measuring its brightness.
5) You can also measure how far away a star is using something called PARALLAX.
6) It's VERY HARD to measure how far away a star is.
7) So if someone says they can measure it spot on, they're lying.

Atmosphere and Light Pollution Cause Problems

The atmosphere is the layer of gas around the Earth.

1) The ATMOSPHERE ABSORBS some light from space.
2) This makes it hard for us to see things in space.
3) LIGHT POLLUTION is the light coming from things like street lamps.
4) Light pollution makes it hard to see DIM objects like some stars:

No light pollution = LOTS OF STARS

Lots of light pollution = NO STARS

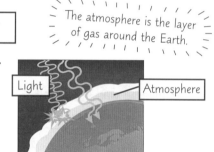

We See Stars as They Were In The Past

1) Light travels really fast in space.
2) BUT space is VERY, VERY BIG.
3) So it can take A LONG TIME for light to reach us on Earth.
4) We see stars how they looked when light first left them.
5) So we see them as they were in the PAST.

Light from the Sun takes 8 mins to reach Earth.

So we see it how it looked 8 mins ago.

Practice Questions

1) What do stars give out?
2) Why is light pollution bad when looking into space?
3) True or false: we see stars as they will look in the future.

The Life of the Universe

The Universe has been around for about 14 thousand million years. That's a long time...

The Universe Started With a Big Bang

1) Other galaxies are MOVING AWAY from us.
2) This shows that the Universe is getting BIGGER.
3) All the galaxies are moving apart really quickly from a single point.
4) Scientists think a big explosion called the BIG BANG got them going:

A theory is an idea that helps to explain something.

THE BIG BANG THEORY:

1) All the stuff in the Universe was squashed into a tiny space.
2) 14 thousand million years ago it exploded and started getting bigger. This was the Big Bang.
3) The Universe is still getting bigger today.

Distances in Space are Hard To Measure

It's hard to tell IF and WHEN the Universe will END.

It's hard for two reasons:

1) It's tricky to measure the VERY LARGE DISTANCES in space.
2) It's tricky to look at how things MOVE in space.

Practice Questions

1) Are galaxies moving apart or getting closer together?
2) How long ago was the Big Bang?
3) Give two reasons why it's hard to tell if and when the Universe will end.

Module P1 — The Earth in the Universe

The Changing Earth

Rocks can tell us a lot about the Earth.

Rocks Show that the Earth is Changing

1) The Earth has CHANGED over time.
2) NEW rock and new mountains have been made.
3) Without new rock, continents would be worn down and the Earth would be SMOOTH.
4) CHANGES in ROCKS show us that the Earth has changed.
5) Rocks can change in different ways:

Continents are big chunks of land like Africa or America.

EROSION:

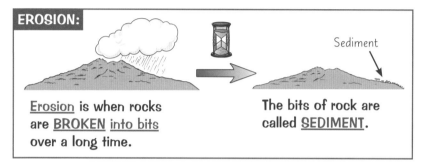

Erosion is when rocks are BROKEN into bits over a long time.

The bits of rock are called SEDIMENT.

SEDIMENTATION:

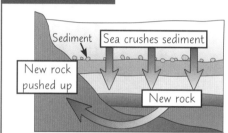

1) Sediment is CRUSHED together under the sea.
2) After thousands of years the sediment becomes new rock.
3) The rock is PUSHED UP to the surface of the Earth.

FOSSILS:

Rock builds up around the fossil over time.

1) FOSSILS are bits of really old animals and plants found in rocks.
2) The animals and plants couldn't have DUG themselves into the middle of rocks.
3) The rocks must have BUILT UP around them.

FOLDING:

Sometimes rocks are squeezed together so much that they FOLD.

The Earth is at least 4000 million years old

Earth's oldest rock — 4000 million years old.

1) The Earth is at least as old as its oldest ROCKS.
2) The oldest rocks found so far are about 4 thousand million years old.
3) So the Earth must be at least 4 thousand million years old.

Practice Questions

1) What is erosion?
2) How old are the oldest rocks found so far on Earth?

Module P1 — The Earth in the Universe

Wegener's Theory of Continental Drift

A <u>theory</u> is an <u>idea</u> that helps to <u>explain</u> something.
So Wegener's theory of <u>continental drift</u> is just Wegener's idea that the <u>continents can move</u>.

Wegener's Theory of Continental Drift

1) A man called <u>WEGENER</u> had a <u>theory</u> that the <u>continents can move</u>.
2) He called his theory <u>CONTINENTAL DRIFT</u>.

This is how the continents look now:

These are <u>CONTINENTS</u>. Wegener thought they were <u>JOINED TOGETHER</u> like this millions of years ago.

Wegener thought that the Earth was made of <u>CHUNKS</u> that <u>SPLIT APART</u>. When the chunks moved they took the <u>continents</u> with them.

Wegener thought <u>MOUNTAINS</u> were made by <u>CHUNKS CRASHING TOGETHER</u>.

mountain building

EVIDENCE for Continental Drift

- Wegener had seen that the <u>continents</u> would fit together like a <u>JIGSAW</u>.
- He also found <u>MATCHING FOSSILS</u> in <u>different continents</u>.
- There are <u>MATCHING LAYERS</u> in the <u>rocks</u> on different continents too.

Nobody Believed Wegener For a Long Time

Most scientists <u>DIDN'T BELIEVE</u> Wegener's theory at first. Here's why...

1) Wegener <u>didn't</u> have much <u>EVIDENCE</u> that he was right.
2) No one could <u>SEE</u> that the <u>continents were moving</u>.
3) Most people had <u>SIMPLER IDEAS</u> and explanations.
4) Wegener <u>wasn't</u> a <u>GEOLOGIST</u>.

You're no geologist, Wegener!

Your ideas are rubbish!

A <u>geologist</u> is a type of scientist who studies the Earth.

But in the end <u>NEW EVIDENCE</u> showed that Wegener was <u>RIGHT</u> all along:

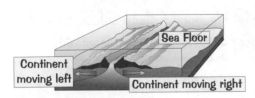

- Scientists found the <u>SEA FLOOR</u> was <u>SPREADING APART</u>.
- This meant the <u>CONTINENTS WERE MOVING</u>.

Practice Questions

1) True or false: Wegener thought that the Earth was made of <u>chunks</u> that <u>split apart</u>.
2) Give one reason why people <u>didn't believe</u> Wegener's theory at first.

Module P1 — The Earth in the Universe

The Structure of the Earth

We can tell what goes on inside the Earth by looking at what happens on the surface.

The Earth Has a Crust, Mantle and Core

The Earth is made up of different layers.

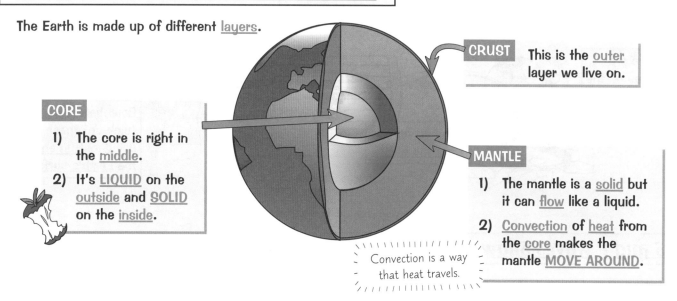

CRUST This is the outer layer we live on.

CORE
1) The core is right in the middle.
2) It's LIQUID on the outside and SOLID on the inside.

MANTLE
1) The mantle is a solid but it can flow like a liquid.
2) Convection of heat from the core makes the mantle MOVE AROUND.

Convection is a way that heat travels.

Tectonic Plates Float on the Mantle

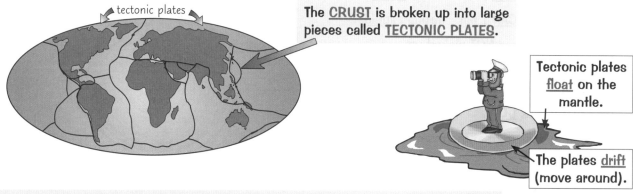

tectonic plates

The CRUST is broken up into large pieces called TECTONIC PLATES.

Tectonic plates float on the mantle.

The plates drift (move around).

1) Tectonic plates move because of CONVECTION in the mantle.
2) The movement of the plates creates EARTHQUAKES, MOUNTAINS and VOLCANOES.
3) These things mostly happen at the EDGES of tectonic plates.

It's the MOVING MANTLE that makes the sea floor SPREAD:

Sea floors spread by a few centimetres every year.

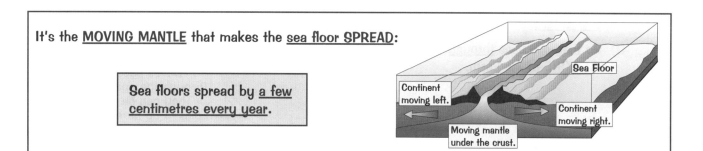

Sea Floor
Continent moving left.
Continent moving right.
Moving mantle under the crust.

Practice Questions

1) What is the outer layer of the Earth called?
2) Name three things that happen at the edges of tectonic plates.

Module P1 — The Earth in the Universe

Seismic Waves

Waves from earthquakes help scientists find out what's inside the Earth.

Earthquakes Cause Waves

Earthquakes cause waves called SEISMIC WAVES.

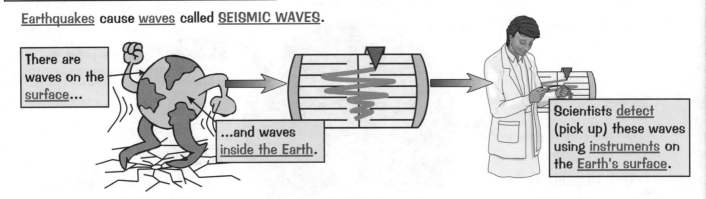

S and P Waves Show What the Earth is Made of

S-WAVES and P-WAVES are two types of seismic wave. They are both caused by earthquakes.

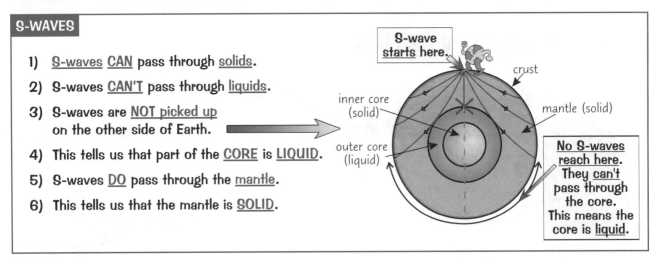

S-WAVES

1) S-waves CAN pass through solids.
2) S-waves CAN'T pass through liquids.
3) S-waves are NOT picked up on the other side of Earth.
4) This tells us that part of the CORE is LIQUID.
5) S-waves DO pass through the mantle.
6) This tells us that the mantle is SOLID.

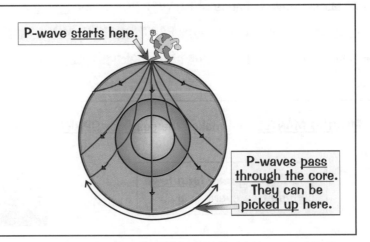

P-WAVES

1) P-waves can pass through liquids AND solids.
2) P-waves can be picked up everywhere.
3) So they can pass through the CORE of the Earth.

Practice Questions

1) True or false: earthquakes cause seismic waves.
2) Which type of wave can pass through liquids, P-waves or S-waves?

Module P1 — The Earth in the Universe

Waves — The Basics

You have to know what all the words to do with waves mean.

Waves Have Frequency, Amplitude and Wavelength

1) Waves are caused by something VIBRATING (shaking).
2) Waves TRANSFER (move) ENERGY from place to place.
3) The energy moves in the SAME DIRECTION that the wave travels.
4) Waves DO NOT transfer MATTER.
5) All waves have FREQUENCY, AMPLITUDE and WAVELENGTH:

Matter is just stuff like water and air.

FREQUENCY

1) This is the how many whole waves pass a certain point EVERY SECOND.
2) OR the number of waves made each second.
3) Frequency is measured in hertz (Hz).
4) 1 Hz is 1 wave per second.

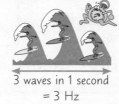
3 waves in 1 second = 3 Hz

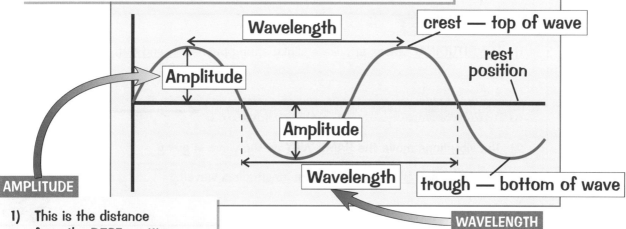

AMPLITUDE

1) This is the distance from the REST position to a CREST or TROUGH.
2) Waves with bigger amplitude carry more energy.

WAVELENGTH

1) This is the length of a FULL CYCLE of the wave.
2) For example, measuring the distance from crest to crest gives you the wavelength.

Speed Is How Fast It Goes

You can work out the distance a wave has travelled using this FORMULA:

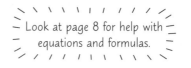

distance = speed × time
metres (m) metres per second (m/s) seconds (s)

EXAMPLE:
A wave is travelling at 5 m/s. How far does it travel in 20 s?

ANSWER: Distance = speed × time = 5 m/s × 20 s = 100 m

Look at page 8 for help with equations and formulas.

Practice Questions

1) True or false: waves transfer energy from place to place.
2) What is frequency measured in?
3) A wave is travelling at 3000 m/s. Find the distance it travels in 60 s.

Module P1 — The Earth in the Universe

Waves — The Basics

More on waves coming right up.

Waves Can Be Transverse or Longitudinal

TRANSVERSE WAVES

1) TRANSVERSE waves are like a slinky spring wiggled from side to side:

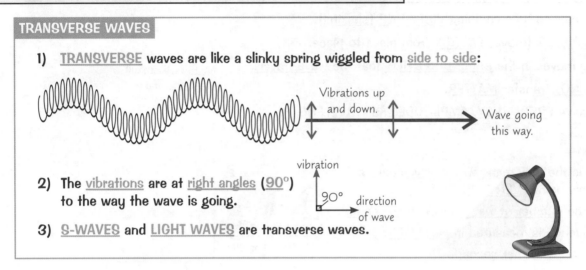

2) The vibrations are at right angles (90°) to the way the wave is going.

3) S-WAVES and LIGHT WAVES are transverse waves.

LONGITUDINAL WAVES

1) LONGITUDINAL waves are like a slinky spring pushed in and out:

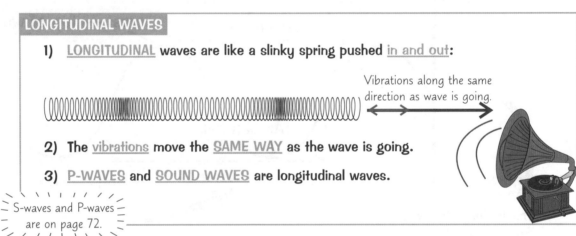

2) The vibrations move the SAME WAY as the wave is going.

3) P-WAVES and SOUND WAVES are longitudinal waves.

S-waves and P-waves are on page 72.

Wave Speed = Frequency × Wavelength

This equation lets you work out the SPEED OF A WAVE:

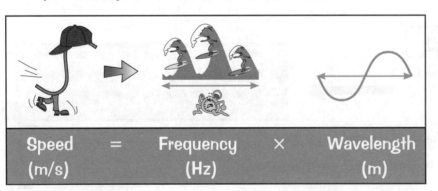

| Speed (m/s) | = | Frequency (Hz) | × | Wavelength (m) |

EXAMPLE:
A wave has a frequency of 0.5 Hz.
Its wavelength is 0.9 m.
What is its speed?

ANSWER:
Speed = frequency × wavelength
= 0.5 Hz × 0.9 m
= 0.45 m/s

Practice Questions

1) Is light a transverse or longitudinal wave?
2) A sound wave has a frequency of 16 000 Hz and a wavelength of 0.2 m. Find its speed.

Module P1 — The Earth in the Universe

Module P2 — Radiation and Life

Electromagnetic Radiation

Light, X-rays and microwaves are all types of electromagnetic radiation.

Light is One Sort of Electromagnetic Radiation

1) There are SEVEN different sorts of ELECTROMAGNETIC RADIATION.
2) Together they make up the ELECTROMAGNETIC SPECTRUM:

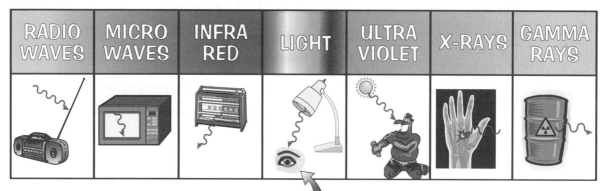

3) LIGHT is just electromagnetic radiation that we can SEE.

Electromagnetic Radiation Carries Energy

1) All electromagnetic radiation carries ENERGY from one place to another.
2) The energy is carried by PHOTONS.
3) A photon is a tiny packet of ENERGY.
4) Electromagnetic radiation can be called electromagnetic WAVES. It can be drawn like this:

Radio Waves Have the Lowest Energy

1) Different sorts of electromagnetic radiation have different FREQUENCIES and ENERGIES.
2) The FREQUENCY of electromagnetic waves INCREASES as you go along the spectrum from left to right.

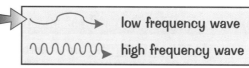

3) The AMOUNT of ENERGY that each photon carries INCREASES as you go along the spectrum.

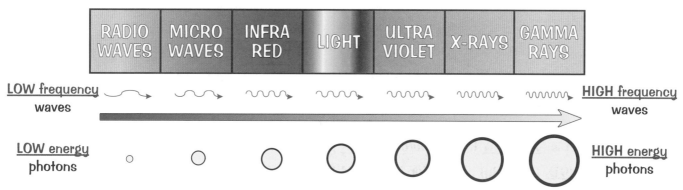

Practice Questions
1) What is the name of the packets that carry energy?
2) Which electromagnetic waves have the lowest energy?

Electromagnetic Radiation and Energy

All objects give out some sort of electromagnetic radiation — you, the Sun, a light bulb, your porridge...

Radiation Comes From a Source

1) Electromagnetic radiation always comes FROM something. We say it is EMITTED from a SOURCE.
2) It travels through SPACE at 300 000 kilometres per second. This is VERY FAST.
3) When it HITS something it gets REFLECTED, TRANSMITTED or ABSORBED.

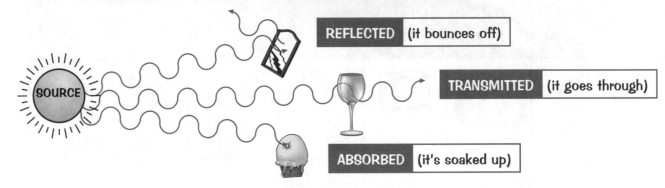

REFLECTED (it bounces off)

TRANSMITTED (it goes through)

ABSORBED (it's soaked up)

Hot Sources = High Frequency Radiation

REALLY HOT sources give off HIGH FREQUENCY radiation.

LESS HOT sources give off LOWER FREQUENCY radiation.

Detectors Absorb Radiation

1) Something that ABSORBS electromagnetic radiation is called a DETECTOR.
2) When radiation is absorbed, it passes ENERGY to the detector.
3) The AMOUNT of energy depends on two things:

1) HOW MANY photons are absorbed

If LOTS of PHOTONS are absorbed, LOTS of ENERGY is absorbed.

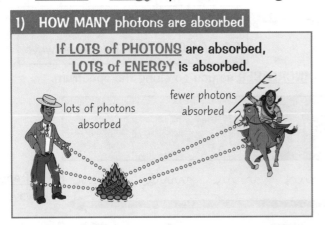

lots of photons absorbed

fewer photons absorbed

2) The ENERGY of each photon

If EACH PHOTON has LOTS of ENERGY, LOTS of ENERGY is absorbed overall.

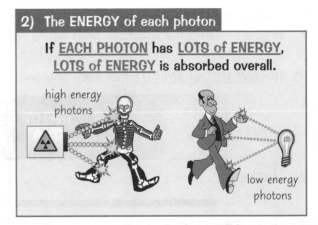

high energy photons

low energy photons

4) If an object is FAR AWAY from a source, FEWER photons will reach it and LESS energy will be ABSORBED.
5) The STRENGTH (intensity) of radiation is HOW MUCH ENERGY hits a surface EACH SECOND.

Practice Questions

1) How fast does radiation travel through space?
2) Which source gives off higher frequency radiation, the Sun or a radiator?

Module P2 — Radiation and Life

Ionising Radiation

Ultraviolet radiation, X-rays and gamma rays can harm people.

Some Electromagnetic Radiation is Ionising

1) Ultraviolet radiation, X-rays and gamma rays are called IONISING RADIATION.
2) They have photons with really HIGH ENERGIES.
3) If they hit an atom or molecule they can KNOCK OFF an electron.
4) This makes an ION.

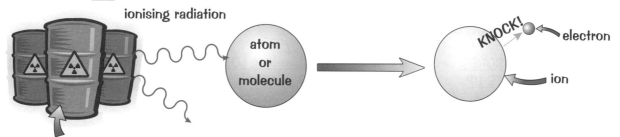

5) RADIOACTIVE materials give off (emit) gamma radiation all the time.

Ionising Radiation can be Harmful

Ionising radiation damages your CELLS.

LOTS of ionising radiation will cause cell DEATH or CANCER.

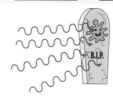

You Need Protection From Dangerous Radiation

1) Sunscreen and clothes protect us from ULTRAVIOLET RADIATION.
2) They ABSORB the radiation so it doesn't reach our skin.

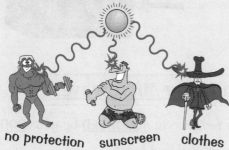

no protection — SUNBURNT
sunscreen / clothes — NOT SUNBURNT

1) X-RAYS can pass through FLESH.
2) But they are absorbed by DENSE materials, like BONE.
3) This means they can be used to take pictures of bones.

4) X-RAYS are also absorbed by LEAD and CONCRETE.
5) So they are used to protect hospital staff from X-rays.
6) X-rays are also used to check bags at airports.

Don't panic. It's not dangerous to have an X-ray done once in a while — just don't make a habit of it.

Practice Questions

1) Name three types of ionising radiation.
2) What does ionising radiation do to your cells?
3) Name three materials that absorb X-rays.

Module P2 — Radiation and Life

Microwaves

Microwaves are useful for heating food and making phone calls.

Some Electromagnetic Radiation Causes Heating

1) If something ABSORBS electromagnetic radiation, it HEATS UP.
2) If CELLS in your body absorb radiation they heat up.
3) Absorbing LOTS of radiation makes cells really HOT.
4) Absorbing radiation for a LONG TIME also makes cells really hot.
5) This DAMAGES the cells.

Microwave Ovens Give Out Microwaves

1) Microwaves are absorbed by WATER molecules. This heats them up.

2) In microwave ovens, microwaves heat up the water molecules in FOOD.
3) This heats the FOOD up.

4) The metal case and door screen REFLECT or ABSORB the microwaves.
5) This stops the microwaves getting OUT and heating YOU up.

Some People Say There are Health Risks with Using Microwaves

1) Microwaves carry calls between MOBILE PHONES and mobile PHONE MASTS.
2) Some microwave radiation is ABSORBED by your BODY.
3) Some people are worried that using mobile phones might cause HARM.
4) But there ISN'T much evidence that mobiles are dangerous.

Practice Questions

1) What molecules absorb microwaves in a microwave oven?
2) True or false: mobile phones are very bad for your health.

Module P2 — Radiation and Life

Electromagnetic Radiation and the Atmosphere

The atmosphere is the air round the Earth. It keeps us warm by trapping heat.

Some Radiation can Pass Through the Atmosphere

1) The Earth is surrounded by an ATMOSPHERE made of gases.

2) THE SUN gives out electromagnetic radiation.

3) Some electromagnetic radiation PASSES THROUGH the atmosphere.

4) This electromagnetic radiation is ABSORBED — WARMING the Earth.

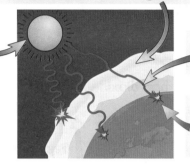

The Greenhouse Effect Helps to Keep the Earth Warm

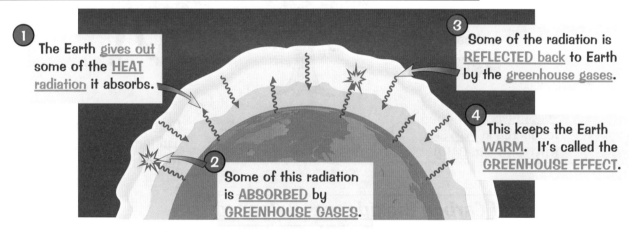

① The Earth gives out some of the HEAT radiation it absorbs.

② Some of this radiation is ABSORBED by GREENHOUSE GASES.

③ Some of the radiation is REFLECTED back to Earth by the greenhouse gases.

④ This keeps the Earth WARM. It's called the GREENHOUSE EFFECT.

1) GREENHOUSE GASES are the gases in the atmosphere that can ABSORB and REFLECT HEAT RADIATION.
2) There are only SMALL amounts of greenhouse gases in our atmosphere.
3) CARBON DIOXIDE, WATER VAPOUR and METHANE are three greenhouse gases.

The Ozone Layer Protects Us

1) The OZONE LAYER is a layer of the atmosphere.
2) It ABSORBS ULTRAVIOLET RADIATION from the Sun.

3) The ozone layer PROTECTS LIFE on Earth from harmful ultraviolet radiation.

Practice Questions

1) Name three greenhouse gases.
2) True or false: the ozone layer absorbs ultraviolet radiation from the Sun.

Module P2 — Radiation and Life

The Carbon Cycle

All the carbon on Earth moves in a big cycle — the carbon cycle.

The Carbon Cycle Shows How Carbon Moves Around

1) CARBON is found in loads of things and it can move from place to place.
2) This is shown using the CARBON CYCLE.

Carbon mostly moves around in the air as the gas carbon dioxide.

PHOTOSYNTHESIS in plants TAKES carbon dioxide from the air.

RESPIRATION in plants and animals ADDS carbon dioxide to the air.

Burning fossil fuels and plants ADDS carbon dioxide to the air.

Fossil fuels are burnt for energy in POWER STATIONS and CARS.

Dead plants and animals can turn into FOSSIL FUELS (oil, gas and coal).

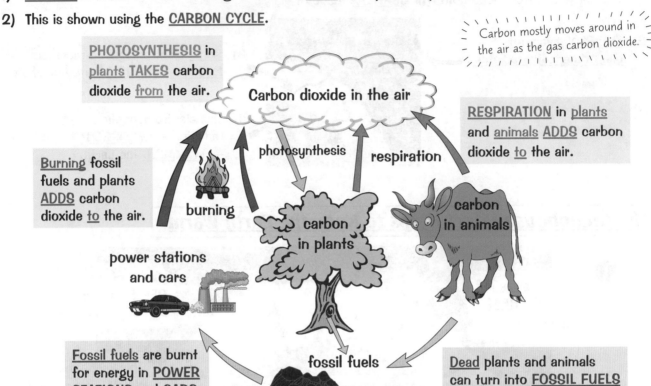

The Amount of Carbon Dioxide was Balanced

The AMOUNT of carbon dioxide in the air has been the SAME (balanced) for thousands of years.

 =

The AMOUNT of carbon dioxide TAKEN from the air.   The AMOUNT of carbon dioxide ADDED to the air.

Humans are Upsetting the Balance

1) For the past 200 years, the amount of carbon dioxide in the air has been INCREASING.
2) It's increasing because of TWO things:

1. More TREES are being CUT DOWN and BURNT to clear land. This ADDS carbon dioxide to the air.

2. More FOSSIL FUELS are being BURNT for energy. This ADDS carbon dioxide to the air.

Practice Questions

1) Does photosynthesis add carbon dioxide to the air or take carbon dioxide from the air?
2) Do respiration and burning add carbon dioxide to the air or take carbon dioxide from the air?

Module P2 — Radiation and Life

Global Warming and Climate Change

We need the greenhouse effect to keep the Earth warm — but we don't want to upset it and overheat.

Upsetting the Greenhouse Effect Has Caused Global Warming

1) The amount of carbon dioxide in the air has INCREASED.

2) This has upset the greenhouse effect.

3) Upsetting the greenhouse effect has made temperatures RISE. This is GLOBAL WARMING.

4) It has caused OTHER changes to the weather too. This is CLIMATE CHANGE.

Global Warming Could Have Bad Effects

1. Rising Sea Level

1) As the sea gets WARMER it GETS BIGGER.
2) This makes the sea level RISE.
3) As the temperature gets warmer ice MELTS.
4) This also makes the sea level RISE.
5) If the sea level rises some places could FLOOD.

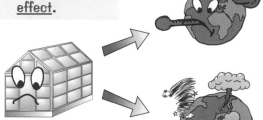

2. Worse Weather

1) As it gets WARMER, some places will get less and less rain. This could cause a DROUGHT.

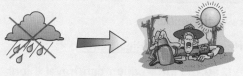

2) Some places could get more HURRICANES. This could cause FLOODS and other problems.

3. Food Crops Not Growing

1) As it gets warmer, some places will be too dry to grow food.
2) Other places might be too wet to grow food.
3) So there might not be enough FOOD.

Practice Questions

1) What will happen to weather because of global warming?
2) What will happen to sea level because of global warming?

Electromagnetic Waves and Communication

We use electromagnetic waves for communication — like phones, TV and radio.

Information is Mixed Onto Electromagnetic Waves

1) ELECTROMAGNETIC WAVES can be used to carry INFORMATION.
2) Information is things like your voice on a phone or the pictures of a TV programme.
3) Information is sent as a SIGNAL.
4) To make a signal, information is MIXED onto electromagnetic waves.

wave + information → signal

Radio Waves and Microwaves are Used for Radio and TV

1) Most radio waves and microwaves DON'T get absorbed by the Earth's atmosphere.
2) So they are good at carrying information a long way.

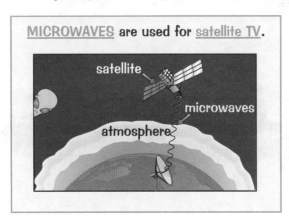

MICROWAVES are used for satellite TV.

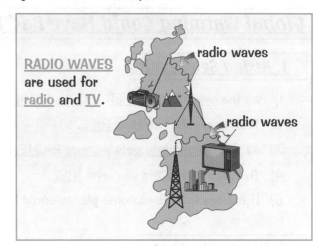

RADIO WAVES are used for radio and TV.

Infrared and Light are Used in Optical Fibres

1) OPTICAL FIBRES are just cables that have a glass core in the middle.
2) LIGHT and INFRARED radiation are used to carry information along optical fibres.
3) The glass doesn't ABSORB much of the light or infrared, so the information can travel a long way.

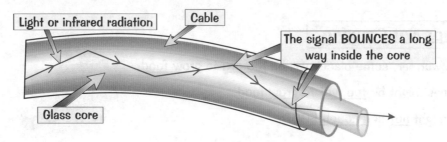

Practice Questions

1) Do radio waves and microwaves get absorbed by the Earth's atmosphere?
2) Are radio waves used to carry TV programmes?
3) What two sorts of radiation are used to carry information along optical fibres?

Module P2 — Radiation and Life

Analogue and Digital Signals

Information like sound or pictures is sent as a signal. There are two different types of signal.

Analogue Signals can have Any Number

1) Some signals are called ANALOGUE SIGNALS.
2) An ANalogue signal can have ANy number.
3) We say it can vary CONTINUOUSLY.

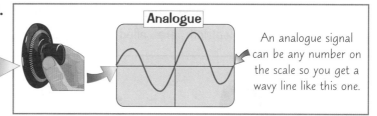

An analogue signal can be any number on the scale so you get a wavy line like this one.

Digital Signals are Either Off or On

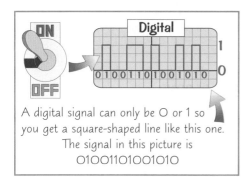

A digital signal can only be 0 or 1 so you get a square-shaped line like this one. The signal in this picture is 01001101001010

1) Some signals are called DIGITAL SIGNALS.
2) Digital signals can only be 0 or 1.
3) They are made by switching the electromagnetic wave OFF and ON.
4) This creates small bursts of waves called PULSES.

OFF = no wave pulse = 0   ON = wave pulse = 1

5) A digital RECEIVER will use the pulses to work out (decode) the signal.

Digital Signals are Clearer

1) NOISE is something that messes up a signal.
2) Signals pick up noise as they travel.
3) Noise is LESS of a problem with digital signals.
4) So digital signals are CLEARER.

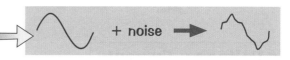

Computers Use Digital Signals

1) Digital signals can carry PICTURES and SOUND.
2) Computers can STORE and USE these signals.
3) The amount of INFORMATION used to store pictures or sounds is measured in BYTES.
4) MORE information means CLEARER images or sounds.

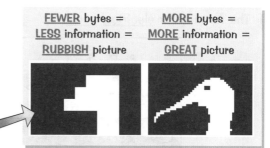

FEWER bytes = LESS information = RUBBISH picture
MORE bytes = MORE information = GREAT picture

Practice Questions

1) True or false: analogue signals can only be 0 or 1.
2) Is noise more or less of a problem with digital signals?
3) What do bytes measure?

Module P2 — Radiation and Life

Electrical Energy

I hope you're feeling lively — this whole module is about energy.

Electrical Energy Comes from a Power Supply

1) Electricity is just a special type of energy. It's called ELECTRICAL ENERGY.
2) Things that use electricity are called COMPONENTS, DEVICES or APPLIANCES.

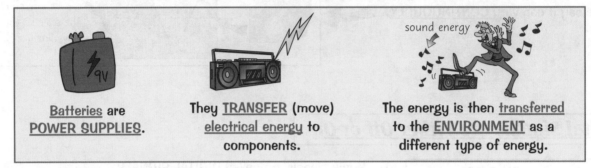

Batteries are POWER SUPPLIES.

They TRANSFER (move) electrical energy to components.

The energy is then transferred to the ENVIRONMENT as a different type of energy.

Power is How Fast Energy is Transferred

The POWER of an appliance tells you HOW MUCH energy it transfers EACH SECOND.

UNITS

POWER is measured in WATTS, W, or KILOWATTS, kW.
1 kW = 1000 W.

EXAMPLE

A 100 W bulb transfers 100 J of energy every second.

A 2 kW kettle transfers 2000 J of energy every second.

1) The kettle transfers more energy every second than the bulb.
2) So, the kettle has a HIGHER POWER than the bulb.

Power is Voltage x Current

1) Something with a HIGH POWER transfers a LOT of energy in a SHORT TIME.
2) This energy comes from the CURRENT flowing through it.
3) Current is the MOVEMENT OF ELECTRICITY around the circuit.
4) The formula for electrical power is:

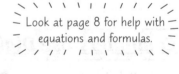

Look at page 8 for help with equations and formulas.

POWER (in W) = VOLTAGE (in V) × CURRENT (in A)

Practice Questions

1) True or false: batteries are power supplies.
2) An iron has a power of 1000 W. How much energy does it transfer every second?
3) A toy robot has a 9 V battery and a current of 17 A. Find its power.

Electrical Energy

The formulas on this page are handy for working out energy bills.

Energy Transferred = Power x Time

1) Appliances can TRANSFER ENERGY.

An appliance is anything that uses electricity — like a TV or an oven.

2) You can find the amount of energy transferred using this formula:

ENERGY TRANSFERRED (in J) = POWER (in W) × TIME (in s)

EXAMPLE:
A kettle has a power of 2500 W. It's left on for 300 s. How much energy is transferred?

ANSWER:
Energy = Power × Time
= 2500 W × 300 s
= 750 000 J

Kilowatt-hours are UNITS of Energy

1) ENERGY is usually measured in JOULES, J.
2) But one joule is a TINY amount of energy.
3) ELECTRICITY METERS measure the energy you use in units of KILOWATT-HOURS (kWh) instead.

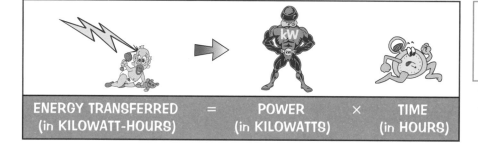

ENERGY TRANSFERRED (in KILOWATT-HOURS) = POWER (in KILOWATTS) × TIME (in HOURS)

EXAMPLE:
A tumble drier has a power of 4 kW. It's on for 1.5 hours. How much energy is transferred?

ANSWER:
Energy = Power × Time
= 4 kW × 1.5 h = 6 kWh

Electricity Costs Money

You can work out how much electricity COSTS with this formula:

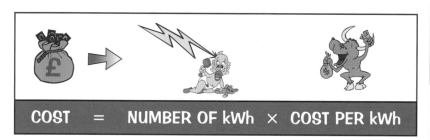

COST = NUMBER OF kWh × COST PER kWh

EXAMPLE:
A light bulb uses 8 kWh in one month. 1 kWh costs 10p. How much will it cost to use the light bulb for one month?

ANSWER:
Cost = number of kWh × cost per kWh
= 8 kWh × 10p = 80p

Practice Questions

1) What is the formula for energy transferred?
2) a) A 2 kW popcorn maker is left on for 11 hours. How much energy is transferred?
   b) 1 kWh costs 20p. How much does it cost to leave the popcorn maker running for 11 hours?

Module P3 — Sustainable Energy

Efficiency

Not all transferred energy is useful — some always gets wasted.

Energy Can't be Lost or Gained

1) When you use electricity, no energy is lost.
2) It's just CHANGED into a different form:

We say the energy is conserved — it's not used up.

EXAMPLE:

A TV converts electrical energy to LIGHT, SOUND and HEAT energy.

ELECTRICITY →

→ LIGHT and SOUND
This is USEFUL energy (the type of energy we want).

→ HEAT
This is WASTED energy.

3) For an appliance, you might only know the TOTAL energy and the amount of WASTED energy...
4) ... but you can use this equation to work out the USEFUL energy:

USEFUL energy = TOTAL energy − WASTED energy

The total energy is all the energy you've put in.

Efficient Appliances Waste Less Energy

1) EFFICIENT appliances change MOST energy to USEFUL energy.
2) This also means they don't WASTE MUCH ENERGY.
3) You can use this equation to find efficiency:

$$\text{EFFICIENCY} = \frac{\text{USEFUL ENERGY}}{\text{TOTAL ENERGY}}$$

EXAMPLE: A light bulb uses 20 000 J of energy. 1000 J is given off as useful light energy. What is the efficiency of the bulb?

ANSWER: $\text{EFFICIENCY} = \dfrac{\text{USEFUL energy}}{\text{TOTAL energy}} = \dfrac{1000}{20\,000} = \underline{0.05}$

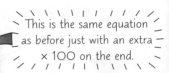

heat 19 000 J
light 1000 J
20 000 J

You Can Give Efficiency as a Percentage

You can use this equation to give efficiency as a percentage:

$$\text{EFFICIENCY} = \frac{\text{USEFUL ENERGY}}{\text{TOTAL ENERGY}} \times 100$$

This is the same equation as before just with an extra × 100 on the end.

So, for the example above, the efficiency as a percentage would be:

$\text{EFFICIENCY} = \dfrac{\text{USEFUL energy}}{\text{TOTAL energy}} \times 100 = \dfrac{1000}{20\,000} \times 100 = \underline{5\%}$

This light bulb isn't very efficient.

Practice Questions

1) Name **one** type of useful energy made by a TV.
2) A robot is used to tickle a mouse. 300 000 J are used to tickle the mouse but 30 000 J are wasted as heat. How efficient is the robot? Give your answer as a percentage.

Module P3 — Sustainable Energy

Sankey Diagrams

Sankey diagrams are just a fancy way of showing where energy goes. Phew.

Sankey Diagrams Show Where Energy Goes

SANKEY DIAGRAMS show how much of the energy that goes INTO an appliance is turned into USEFUL energy and how much is WASTED energy.

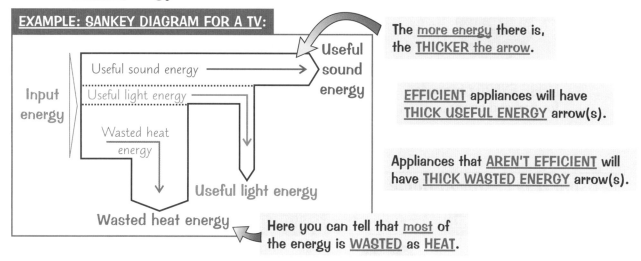

Power Stations Are Not Very Efficient

1) Power stations GENERATE (make) electricity.
2) But lots of energy is WASTED as HEAT and NOISE.
3) The electricity is sent from the power station to people's homes — this is called DISTRIBUTION.
4) Some energy is wasted as HEAT during distribution.
5) This can be shown in a Sankey diagram like the one here:

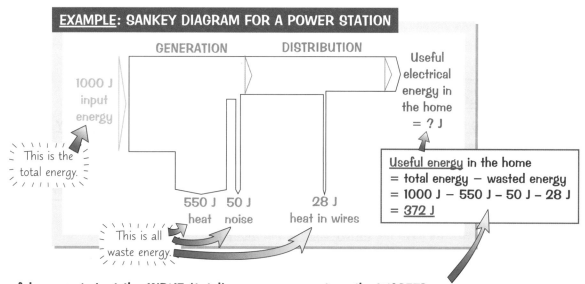

6) The useful energy is just the INPUT (total) energy arrow minus the WASTED energy arrows.
7) You can use the equation from the last page to find the EFFICIENCY.

Practice Questions

1) What does a Sankey diagram show?
2) True or false: appliances that aren't efficient have thick useful energy arrows.

Saving Energy

We can all do our bit to save energy — for example, I typieed this put in thee durk.

We Need To Use Less Energy

1. As the population grows, MORE ENERGY is needed for fuel and power.

2. We use FOSSIL FUELS for most of our energy. This is bad because they will eventually run out. They can also harm the ENVIRONMENT.

3. It will help if we REDUCE the amount of energy we use every day.

Using Less Energy At Home, School and Work

STOP HEAT ESCAPING
Insulation, double glazing and draught-proofing all stop heat escaping.

SWITCH OFF
Switch off lights, computers and other appliances when you're not using them.

Energy efficient means most of the energy being used is useful.

GOVERNMENT HELP
The government can get people to save energy by giving them money to buy more efficient boilers. They also make laws to make sure new homes and businesses are energy efficient.

GREEN TRAVEL
Use energy efficient ways to travel like cycling, car-sharing or using public transport.

TURN DOWN HEATING
Turn down the heating and put on warmer clothes instead.

EFFICIENT APPLIANCES
Buy more efficient appliances like energy-saving light bulbs.

Some Changes Cost More than Others

1) ENERGY-SAVING IMPROVEMENTS are things that help you save ENERGY and MONEY.

DOUBLE GLAZING
COSTS: £3000
SAVES: £60 per year

ENERGY SAVING LIGHT BULBS
COSTS: £3
SAVES: £12 per year

2) The BEST improvements save the MOST energy and money.
3) They can also COST money to put in though.
4) You can work out the time taken for an appliance to PAY BACK the money it cost using this equation:

$$\text{TIME TO PAY FOR ITSELF} = \frac{\text{COST}}{\text{AMOUNT SAVED PER YEAR}}$$

Practice Questions

1) What type of fuel do we use most?
2) Give two ways that you could save energy at home.

Module P3 — Sustainable Energy

Energy Sources and Power Stations

Electricity doesn't grow on trees. We get it from different energy sources.

Electricity is an Easy Way to Supply Energy

1) Electricity is very convenient (handy).
2) It can travel LONG DISTANCES through the National Grid cables (see p. 96).
3) It can be used in many DIFFERENT ways.
4) PRIMARY ENERGY sources are used to make electricity.
5) Electricity is a SECONDARY ENERGY source.

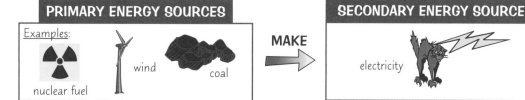

6) Primary sources can be RENEWABLE or NON-RENEWABLE.

RENEWABLE: WILL NEVER RUN OUT	NON-RENEWABLE: WILL ALL RUN OUT ONE DAY
Wind Solar Tides Hydroelectric Biofuels Waves Geothermal	**Fossil fuels** Oil Coal Gas **Nuclear fuels** Uranium Plutonium

Electricity can be Generated Using a Turbine

A turbine is something that turns. It can be turned by using water, wind or steam.

1) Most electricity is made in THERMAL POWER STATIONS:

1 Primary energy sources are used to heat water. This makes STEAM. → **2** The steam drives a TURBINE. → **3** A GENERATOR makes electricity from the movement of the turbine (see p. 96).

2) Many RENEWABLE SOURCES (wind, waves, hydroelectric) drive the turbine directly, WITHOUT steam.

Fossil Fuels Have Good and Bad Points

DISADVANTAGES: (Bad points)

1) They give out carbon dioxide when burnt, which causes climate change (see p. 81).
2) Coal and oil give out sulfur dioxide when burnt, which causes acid rain.
3) Coal mining makes a mess of the landscape.
4) Oil spills cause serious problems for sealife.

ADVANTAGES: (Good points)

1) They produce a lot of energy.
2) They're pretty cheap.
3) They don't rely on the weather.
4) We've already got fossil fuel power stations, so we don't need to build new ones.

Practice Questions

1) Why is electricity convenient?
2) Give two advantages of using fossil fuels.

Module P3 — Sustainable Energy

Nuclear Energy

Nuclear power stations can make lots of energy WITHOUT making lots of carbon dioxide.

Nuclear Power Gives Out Lots of Heat Energy

1) A nuclear power station uses nuclear fuel.

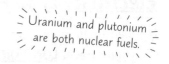
Uranium and plutonium are both nuclear fuels.

2) Here's a block diagram of a nuclear power station:

1	2	3	4
The nuclear fuel releases lots of HEAT energy.	This energy heats water to make STEAM.	The steam turns a TURBINE.	A GENERATOR makes electricity from the movement of the turbine (see p. 96).

Nuclear Power Has Good and Bad Points

ADVANTAGES ✓

1) Nuclear fuel gives out a LOT MORE ENERGY than fossil fuels.
2) Nuclear power stations DON'T produce much carbon dioxide.
3) Nuclear fuel is fairly CHEAP.

DISADVANTAGES ✗

1) Nuclear power stations make RADIOACTIVE WASTE.
2) This waste gives out IONISING RADIATION.
3) Exposure to ionising radiation can DAMAGE living cells.
4) Even more exposure can lead to CANCER or CELL DEATH.
5) The waste is hard to get rid of safely.
6) Nuclear power stations take a very long time to start up.
7) The overall COST of nuclear power is HIGH because it costs a lot to build the power stations.
8) Nuclear fuel will RUN OUT (it's non-renewable).

Exposure means to be around or near to.

Contamination Is Worse Than Irradiation

1) There are TWO ways you can be exposed to radiation:

IRRADIATION
- When you go near radiation but DON'T touch the source.
- The damage stops when you move away so you're only exposed for a short time.

CONTAMINATION
- When you pick up some radioactive waste.
- For example, you might breathe it in, drink affected water or get it on your skin.
- You're still exposed to the radiation when you've left the area.

2) Contamination is WORSE because you're exposed for a LONG TIME, which can cause MORE DAMAGE.

Practice Questions

1) Give one advantage of using nuclear energy.
2) Give two disadvantages of using nuclear energy.
3) Which is worse: contamination or irradiation? Explain your answer.

Module P3 — Sustainable Energy

Wind and Solar Energy

Wind and solar power are renewable sources so they will not run out.

Wind Power

WINDMILLS (wind turbines) are put in OPEN SPACES.

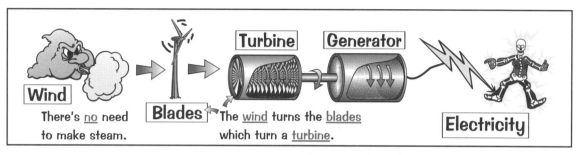

Wind — There's no need to make steam.
Blades — The wind turns the blades which turn a turbine.
Turbine — Generator — Electricity

ADVANTAGES
1) NO pollution or carbon dioxide when you use wind turbines (only a bit when they're made).
2) NO serious damage to the landscape.
3) NO fuel costs and LOW running costs.
4) RENEWABLE — it won't run out.

This is how much it costs to keep the energy source going.

DISADVANTAGES
1) Windmills take up a lot of SPACE, and can SPOIL THE VIEWS.
2) They can be NOISY.
3) They only work when it's WINDY.
4) Start up costs are HIGH.

This is how much it costs to start using the energy.

Solar Cells

1) SOLAR CELLS make electricity from SUNLIGHT.

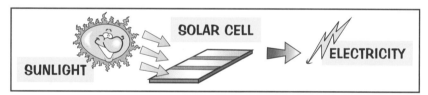

SUNLIGHT → SOLAR CELL → ELECTRICITY

2) Solar power is used in SUNNY places to power SMALL things.

road signs, satellites, single houses

ADVANTAGES
1) NO pollution or carbon dioxide when you use solar panels (only a bit when they're made).
2) Work well in SUNNY places in the DAYTIME.
3) NO fuel costs and LOW running costs.
4) RENEWABLE — it won't run out.

DISADVANTAGES
1) Start up costs are HIGH.
2) They take quite a lot of energy to MAKE.
3) It's hard to connect them to the National Grid (see page 96).
4) They only work when it's SUNNY.

Practice Questions
1) Give one advantage of using wind power.
2) Give one disadvantage of using solar power.

Module P3 — Sustainable Energy

Wave and Tidal Energy

More renewable energy sources here — wave power and tidal power.

Wave Power

Turbines powered by WAVES are put around the COAST.

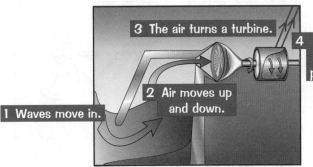

Wave power is only really useful on SMALL ISLANDS.

ADVANTAGES
1) NO pollution or carbon dioxide produced.
2) NO fuel costs and LOW running costs.
3) RENEWABLE — it won't run out.

DISADVANTAGES
1) Wave turbines SPOIL THE VIEW.
2) They're a DANGER to boats.
3) They depend on the WEATHER because waves need wind.
4) Start up costs are HIGH.

Tidal Power

1) TIDAL BARRAGES are big DAMS built across the end of a river.
2) The dams have TURBINES in them.

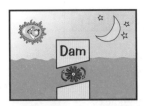

TIDES are caused by the GRAVITY of the Sun and the Moon.

As the TIDE comes in the water rises and TURNS the turbines.

The water is LET OUT through the turbines making electricity.

ADVANTAGES
1) NO pollution or carbon dioxide produced.
2) Tides DON'T depend on the weather.
3) They produce LOTS of ENERGY.
4) NO fuel costs and LOW running costs.
5) RENEWABLE — it won't run out.

DISADVANTAGES
1) Tidal barrages SPOIL THE VIEW.
2) They affect BOATS and WILDLIFE habitats.
3) They DON'T WORK when the water level is the SAME either side of the dam.
4) Start up costs are quite HIGH.

Practice Questions
1) Give one disadvantage of using wave power.
2) Give one advantage of using tidal power.

Module P3 — Sustainable Energy

Biofuels and Geothermal Energy

There's more renewable energy in rubbish and rocks.

Biofuels

Biofuels come from PLANTS or rotting WASTE.

They're BURNT like fossil fuels in THERMAL power stations (see p. 89).

They can also be used in some CARS.

ADVANTAGES
1) Biofuels can be made QUICKLY.
2) They are said to be CARBON NEUTRAL.
3) Overall, NO carbon dioxide is added to the atmosphere.

DISADVANTAGES
1) Sometimes FORESTS are CHOPPED DOWN and burnt to make space to grow biofuels.
2) This means wildlife lose their habitats.
3) Rotting and burning forests give off GREENHOUSE GASES.

Geothermal Energy

GEOTHERMAL energy uses heat from UNDERGROUND.

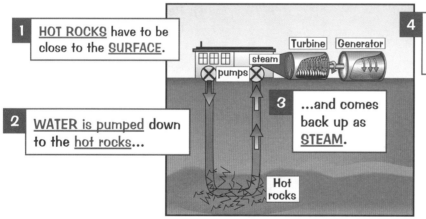

1. HOT ROCKS have to be close to the SURFACE.
2. WATER is pumped down to the hot rocks...
3. ...and comes back up as STEAM.
4. The steam drives a turbine that drives a generator.
5. The generator makes electricity.

ADVANTAGES
1) Not much POLLUTION.
2) LOW running costs.

DISADVANTAGES
1) It costs A LOT to drill down to the hot rocks.
2) There aren't many places where you can easily get geothermal energy.

Practice Questions
1) What are biofuels made from?
2) Give one disadvantage of using biofuels.
3) Give one advantage of using geothermal energy.

Module P3 — Sustainable Energy

Hydroelectricity and Reliable Fuel Supplies

Only one more renewable energy source to go — welcome to the wonderful world of hydroelectricity...

Hydroelectricity

Hydroelectricity uses energy from WATER stored behind a DAM.

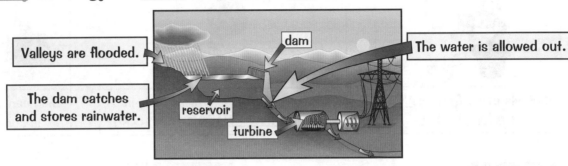

Here's a block diagram:

1. Water is stored above the turbines using a dam.
2. Water is let out and rushes down, turning the turbine.
3. The turbines turn a GENERATOR that makes electricity.

ADVANTAGES

1) No pollution or carbon dioxide when running.
2) Electricity can be made whenever it's needed.
3) It's fairly RELIABLE.
4) NO fuel costs and LOW running costs.

DISADVANTAGES

1) HIGH start up COSTS.
2) Rotting plants in the flooded valley give off CARBON DIOXIDE.
3) Wildlife can lose their HABITATS.
4) Reservoirs behind the dams can look UGLY when they DRY UP.

We Need a Reliable Fuel Supply

1. The UK uses a LOT of electricity. The amount we WANT to use (the DEMAND) is going UP.
2. The BEST energy source would be RELIABLE, CHEAP and ENVIRONMENTALLY-FRIENDLY, and would supply ENOUGH POWER.
3. NO energy source can do all this on its OWN. So we need to use a MIX of energy sources to make sure the UK doesn't run out of electricity.

Practice Questions

1) Give one advantage of using hydroelectricity.
2) Give one disadvantage of using hydroelectricity.
3) Why do we need to use a mix of energy sources in the UK?

Module P3 — Sustainable Energy

Comparing Energy Resources

With all these renewables, who needs smelly fossil fuels? Choosing an energy resource is not that simple...

Think about the Environment as well as the Cost

To COMPARE or CHOOSE BETWEEN energy sources, think about the COST and the ENVIRONMENTAL IMPACT.

COST	ENVIRONMENTAL IMPACT
RENEWABLES usually have the LOWEST RUNNING COSTS.  renewables · fossil fuels	If a FUEL is used, you'll be using up RESOURCES.
	ALL energy sources cause some POLLUTION or WASTE.
 Renewables often need bigger power stations than non-renewables to make the SAME AMOUNT of electricity. renewables · fossil fuels	FOSSIL FUELS produce carbon dioxide which leads to GLOBAL WARMING and CLIMATE CHANGE.
Renewable sources usually have HIGHER START UP costs. renewables · fossil fuels	RENEWABLES can cause NOISE pollution and SPOIL VIEWS.
	NUCLEAR power doesn't produce carbon dioxide but it does make RADIOACTIVE WASTE.

No Energy Source is Perfect

1) You might be given information like this:

	Coal	Gas	Nuclear
Efficiency	36%	50%	38%
Carbon dioxide produced per unit of electricity (g)	920	440	110
Cost of energy per unit (p)	2.5-4.5	2-3	4-7

2) You'll need to pick out the GOOD and BAD points of each energy source from the table.

   EXAMPLE Nuclear power doesn't give off much carbon dioxide, but it costs a LOT per unit of energy. So if you want a CHEAPER fuel, a coal or gas power station would be better.

Practice Questions

1) Do renewable energy sources usually have the highest or the lowest running costs?
2) Name one type of pollution that renewable fuels can cause.

Module P3 — Sustainable Energy

Generators and the National Grid

The energy from power stations goes into a generator and comes out as electricity.

Generators Make Electricity Using Magnets

1) MAINS ELECTRICITY is the electricity we plug into at home.

2) It's made in power stations using GENERATORS.

3) Electricity is made in generators by moving a MAGNET in or near a COIL of wire.

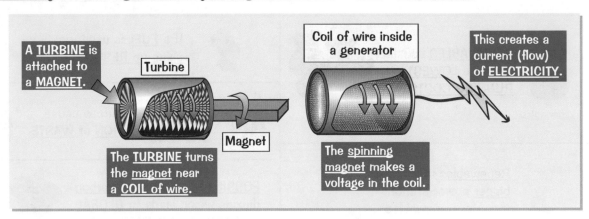

4) Turning the magnet FASTER makes a BIGGER VOLTAGE and CURRENT.

5) But you need to use MORE ENERGY to turn the TURBINE faster, which uses up more PRIMARY FUEL (like coal).

6) So the BIGGER the CURRENT, the MORE FUEL used every second.

Electricity is Carried by the National Grid

1) PYLONS and ELECTRICAL CABLES cover the whole of Britain.

2) This network of electricity is called the NATIONAL GRID.

3) The electricity travels along the CABLES.

4) Electricity could be DISTRIBUTED (sent out) at either a HIGH VOLTAGE or a HIGH CURRENT.

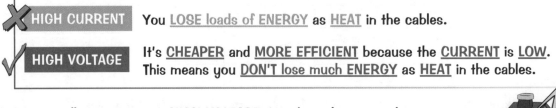

✗	HIGH CURRENT	You LOSE loads of ENERGY as HEAT in the cables.
✓	HIGH VOLTAGE	It's CHEAPER and MORE EFFICIENT because the CURRENT is LOW. This means you DON'T lose much ENERGY as HEAT in the cables.

5) It's actually sent out at a HIGH VOLTAGE to reduce the energy lost.

6) The voltage is reduced to 230 V before it gets to our homes.

7) This is the MAINS SUPPLY VOLTAGE.

Practice Questions

1) True or false: generators use magnets to make mains electricity.

2) What voltage is the mains supply voltage?

Planning

The first part of your controlled assessment is the <u>practical data analysis</u>. You'll be given a <u>hypothesis</u> (see page 2) and you'll need to <u>plan an investigation</u> to <u>test</u> it. Here are a few <u>tips</u> to help you:

Write a Clear Method

1) A method is a <u>STEP-BY-STEP</u> description of everything you would do in an investigation.
2) Someone should be able to <u>REPEAT the investigation</u> using your method, so it needs to be <u>really CLEAR</u>. For example, say <u>EXACTLY how much</u> of a chemical you'd use and <u>how you'd measure it out</u>.
3) You need to say <u>what</u> you'll be <u>measuring</u> and <u>HOW MANY readings</u> you'll take.
4) <u>Repeating</u> the readings and working out the <u>mean</u> (see next page) will make your results more <u>reliable</u>.

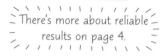
There's more about reliable results on page 4.

Think Carefully About the Equipment You'll Use

1) You need to list all the <u>EQUIPMENT</u> you're going to use.
2) Using the right equipment will give you <u>ACCURATE RESULTS</u> (results near to the real value).
3) For example:

If you need to measure out <u>11 ml</u>, this measuring cylinder would be great. It's the <u>right size</u> and you can <u>see</u> where 11 ml is.

This measuring cylinder isn't as good. It's <u>too big</u> and you <u>can't really see</u> where 11 ml is.

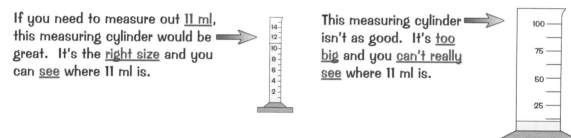

Experiments Must be Safe

1) Part of planning an investigation is making sure that it'll be <u>SAFE</u>.
2) You should always make sure that you think of all the <u>HAZARDS</u> (dangers) that you might come across.
3) Hazards are things like...

microorganisms (e.g. bacteria)   chemicals   fire   electricity

4) You need to come up with ways of <u>reducing the risks</u> from the hazards you've spotted. For example:

- If you're using a <u>Bunsen burner</u>, stand it on a <u>heat-proof mat</u>.
- This will <u>reduce the risk</u> of <u>starting a fire</u>.

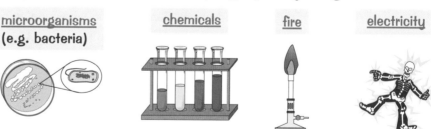

5) You need to <u>write</u> about any <u>hazards</u> and how you'll <u>reduce the risk</u> of them in your plan.

Processing the Data

Once you've planned your experiment you'll carry it out to collect some <u>data</u>.
The next part of your practical data analysis is to <u>process the data</u>.

Data Needs to be Organised

1) Data that's been collected needs to be <u>organised</u>.
2) <u>TABLES</u> are dead useful for <u>organising data</u>.

Temperature (°C)	Repeat 1 (s)	Repeat 2 (s)	Repeat 3 (s)
10	31	30	29
20	22	19	20
30	10	11	11

Make sure each column has a heading
Make sure you give the units

Use a Bit of Maths to Process the Data

1) If you've <u>repeated</u> an investigation you need to work out the <u>MEAN</u> (average).
2) Just <u>ADD TOGETHER</u> the results. Then <u>DIVIDE</u> by the total number of results.

Temperature (°C)	Repeat 1 (s)	Repeat 2 (s)	Repeat 3 (s)	Mean (s)
10	31	30	29	$\frac{(31 + 30 + 29)}{3} = 30$
20	22	19	20	$\frac{(22 + 19 + 20)}{3} = 20.3$
30	10	11	11	$\frac{(10 + 11 + 11)}{3} = 10.7$

Add together the results
Divide by 3 (because there are three results for 10 °C)

If you've got any outliers you can usually ignore them when working out the mean — see page 4 for more on outliers.

Bar Charts are Used When You've Got Categories

1) <u>Charts</u> and <u>graphs</u> are really good ways of showing your results.
2) You can use <u>different</u> graphs and charts depending on what you've been <u>measuring</u>.
3) Bar charts are great when the thing you're measuring has got <u>CATEGORIES</u>.
4) Categories are things like blood type or ice cream flavour. You <u>can't</u> get results <u>in-between categories</u>.
5) There are some <u>rules</u> you need to follow for <u>drawing</u> bar charts...

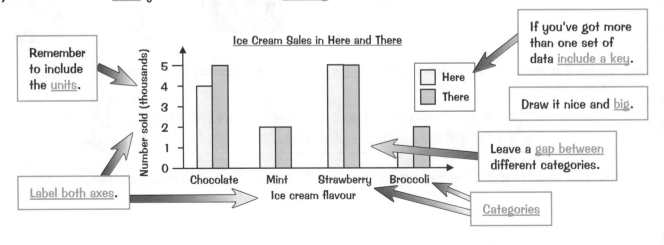

Remember to include the <u>units</u>.

Label both axes.

If you've got more than one set of data <u>include a key</u>.

Draw it nice and <u>big</u>.

Leave a <u>gap between</u> different categories.

Categories

Controlled Assessment — Practical Data Analysis

Processing the Data

Hold up, you haven't finished with data processing yet...

You Need to be Able to Draw Line Graphs

1) If you're measuring something that can have ANY value you should use a LINE graph to show the data.
2) For example, temperatures and people's heights would be shown using a line graph.
3) Here are the rules for drawing line graphs...

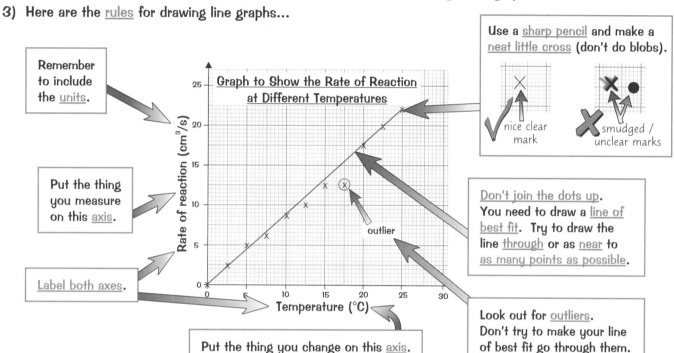

Line Graphs Show Relationships in Data

1) Line graphs are great for showing the relationship (the link) between two things.
2) The relationship is called a CORRELATION. Just look at the pattern on your graph to spot the relationship.

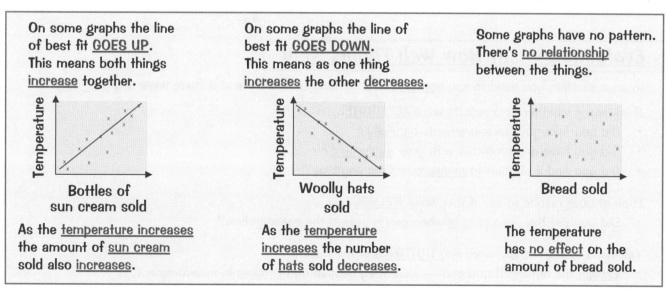

3) Remember, even if there IS a correlation, it DOESN'T always mean that a change in one thing is CAUSING the change in the other (see page 5).

Controlled Assessment — Practical Data Analysis

Conclusion and Evaluation

At the end of your practical data analysis, you'll need to write a conclusion and an evaluation.

A Conclusion is a Summary of What You've Learnt

1) To come to a conclusion, just look at your data and say what pattern you see.

> **EXAMPLE:** Pea plants were grown with different fertilisers. The table below shows how tall the plants grew.
>
>
>
> | Fertiliser | Mean growth / mm |
> |---|---|
> | A | 13.5 |
> | B | 19.5 |
> | C | 5.5 |
>
> **CONCLUSION:** Fertiliser B makes pea plants grow taller than fertiliser A or fertiliser C.

2) You should use the data that's been collected to BACK UP your conclusion.

> For example: Fertiliser B made the pea plants grow more than fertiliser A. The mean growth for fertiliser B was 19.5 mm and only 13.5 mm for fertiliser A.

3) Look back to the hypothesis. If the conclusion AGREES with the hypothesis, it makes us more sure that the hypothesis is right.
4) If the conclusion DOESN'T AGREE with the hypothesis, then it makes us less sure that it's right.
5) You also need to explain your conclusion.

> For example: Fertiliser B has more of the nutrients that the pea plants need to help them grow.

Evaluation — Say How Well Things Went

In an evaluation you need to say how well your investigation went and if there were any problems.

Think about whether your results were ACCURATE...
- Did you take your measurements carefully?
- Did you have any problems with your equipment?
- Did you find it difficult to measure or time anything?

Look at your results to see if they were RELIABLE...
- Did you get the same results when you repeated the investigation?

You need to say if there were any OUTLIERS in the results.
- Explain the outliers if you can — were they caused by mistakes in measurement?
- If you didn't get any outliers then you need to say that you didn't.

Controlled Assessment — Practical Data Analysis

The Case Study

The other half of your controlled assessment is a case study.
It's all about weighing up information about an issue in science.

The Case Study is all About Evaluating Evidence

1) You'll be given some ARTICLES that are all about the same topic.
2) The topic will be one that people disagree about.
3) You need to use the articles to come up with a RESEARCH QUESTION to talk about in your case study.
4) Then you'll need to find more information on the topic and try to answer your question.
5) You'll need to write the whole thing up as a REPORT. Think about the best way to structure it. You get marks for it being in a clear order.

Planning and Research — Collecting Information

Start off by reading the articles that you've been given.
Then do some research into the topic.

- Find lots of SOURCES (other bits of information) on the topic.
  Try text books, newspapers or the internet.
- Choose sources written by people with DIFFERENT VIEWS on the topic, to get both sides of the story.
- Think about how good your sources are and why, and talk about this in your report.
  For example, think about whether they've been written by a scientist or not.
- Make sure you write in your report where you found all of your sources.

Writing About What You've Found

You need to describe what you've found.

- Start off by explaining the science that someone would need to know to understand the topic.
  For example, if the topic is cloning, you'll need to explain what cloning is.
- Then COMPARE what your sources are saying — are they 'for' or 'against' the topic?
  For example, some sources could be for human cloning, others could be against it.
- You need to look at the evidence that the sources have used.
  Say how reliable you think the evidence is and why.

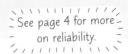
See page 4 for more on reliability.

Writing a Conclusion and Making Recommendations

Finally, you need to write a conclusion and give your own opinion.

- Think about how well the evidence in the sources supports the view that's been given.
- Then write a conclusion — say which viewpoint has the most evidence for it.
- Finally, you need to answer your research question.
  You need to suggest what the best course of action will be.
- Write about the benefits and risks of each course of action.
  Show why the one you picked is the best one.

A course of action is what you decide should be done after thinking about all the evidence for a topic.

Answers

Module B1 — You and Your Genes

Page 9 — Genes, Chromosomes and DNA
1) in the nucleus
2) weight

Page 10 — Genes and Variation
1) alleles
2) one chromosome from each pair
3) false — Children get half their alleles from their mum and half from their dad.

Page 12 — Inheritance and Genetic Diagrams
1) the dominant characteristic
2) a)

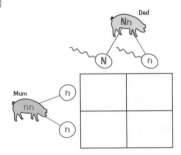

b)

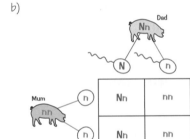

c) 50%

Page 13 — Genetic Diagrams and Sex Chromosomes
1) a) a girl
   b) no
2) XY
3) a girl

Page 14 — Genetic Disorders
1) recessive
2) dominant
3) Any two from: tremors / clumsy / memory loss / mood changes / can't concentrate

Page 15 — Genetic Testing
1) false — Genetic testing can be used on embryos.
2) They can sometimes cause a miscarriage.

Page 16 — Clones
1) the environment
2) asexual reproduction
3) false — Identical twins are clones.

Page 17 — Stem Cells
1) unspecialised
2) embryos, adults
3) any type of cell

Module B2 — Keeping Healthy

Page 18 — Microorganisms and Disease
1) by damaging your cells, by making toxins (poisons)
2) It's warm. It's moist. There's food.
3) You have one bacteria at the start. After 30 minutes you have two bacteria. After 60 minutes you have four bacteria. After 90 minutes you have EIGHT BACTERIA.

Page 19 — The Immune System
1) They digest microorganisms. They make antibodies.
2) chemicals on the surface of a microorganism
3) false — Memory cells remember how to make certain antibodies.

Page 20 — Vaccination
1) false — Dead microorganisms do have antigens.
2) because of differences in their genes

Page 21 — Antibiotics
1) bacteria
2) viruses
3) Take all the antibiotics a doctor gives you. Only use antibiotics when you really need to.

Page 22 — Drug Trials
1) on human cells, on animals
2) a) to check the drug is safe
   b) to check the drug is safe, to check the drug is effective (works)
3) It's not fair to give these patients a fake medicine if the new drug could help them.

Page 23 — The Circulatory System
1) It's two pumps stuck together.
2) Thick, stretchy walls. These help cope with high pressure.
3) Any one from: large hole in the middle — this helps blood flow easily. / Valves — these keep blood flowing in the right direction.

Page 24 — Heart Rate and Blood Pressure
1) your heart rate
2) how hard your blood is pushing against an artery wall
3) Lumps of fat (fatty deposits) build up in the damaged area. The artery becomes blocked. If the artery leads to the heart, it causes a heart attack.

Page 25 — Heart Disease
1) Any three from: poor diet / stress / too much alcohol / smoking / illegal drugs
2) People have a poorer diet. People don't do lots of exercise.
3) by doing a study

Answers

Page 26 — Homeostasis and The Kidneys
1) true
2) receptor, processing centre, effector
3) a) drink, food
   b) Any two from: sweating / breathing / wee / poo

Page 27 — Controlling Water Content
1) by making dilute urine or concentrated urine
2) Any two from: outside temperature / exercise / drinking fluids / eating salt
3) more dilute urine

Module B3 — Life on Earth

Page 28 — Adaptation and Variation
1) a group of organisms that can breed together to make fertile offspring
2) a thick stem, spines and roots that cover a large area
3) a) changes in genes
   b) new features

Page 29 — Natural Selection and Selective Breeding
1) true
2) In natural selection only features that help survival get passed on.

Page 30 — Evolution
1) about 3500 million years ago
2) simple things
3) true

Page 31 — Evolution
1) the fossil record, DNA
2) Developed features aren't caused by genes. This means they can't be passed on to your children.

Page 32 — Biodiversity and Classification
1) the variety of life on Earth
2) Any one from: how similar their DNA is / how similar their physical features are
3) Any one from: A kingdom has lots of types of organisms but a species only has one type of organism. / Organisms in a kingdom have few features in common, but organisms in a species have lots of features in common.

Page 33 — Energy in an Ecosystem
1) blackfly larvae
2) the Sun
3) because so much energy is lost at each stage

Page 34 — Energy in an Ecosystem
1) a) amount of energy lost =
      energy in one stage − energy in previous stage
      10 000 kJ − 1000 kJ = **9000 kJ**
   b) efficiency of energy transfer =
      (energy in one stage ÷ energy in previous stage) × 100
      (1000 ÷ 10 000) × 100 = **10% efficient**

Page 35 — Interactions Between Organisms
1) what is eaten by what
2) Any one from: the environment changes. / Something new turns up, such as something that competes with the species for resources, a predator or a disease. / An organism that it eats becomes extinct.

Page 36 — The Carbon Cycle
1) how carbon is recycled
2) photosynthesis
3) respiration, decomposers, combustion (burning)

Page 37 — The Nitrogen Cycle
1) nitrates in the soil
2) plants and other animals
3) decomposers

Page 38 — Measuring Environmental Change
1) Any one from: temperature / nitrate level / carbon dioxide level
2) organisms that are affected by changes in their environment
3) Any one from: mayfly nymphs (show water pollution) / lichens (show air pollution) / phytoplankton (show water pollution)

Page 39 — Sustainability
1) Sustainability means meeting the needs of people now but without harming the environment. This is so that people in future can still meet their own needs.
2) no
3) false — Monoculture crop production is bad for biodiversity.

Page 40 — Sustainability
1) Any one from: The resources used to make the cardboard can be replaced. / Cardboard is biodegradable.
2) Any one from: by using recycled materials / by using less packaging
3) Any one from: Biodegradable materials still cause pollution. They take a while to break down. / Making and transporting any packaging material uses up energy.

Module C1 — Air Quality

Page 41 — How the Air was Made
1) volcanoes
2) liquid water (the oceans)
3) plants and oceans

Page 42 — The Air Today
1) nitrogen
2) Any four from: carbon dioxide / carbon monoxide / nitrogen oxides / sulfur dioxide / particulates
3) They come from cars, power stations and volcanoes.

Page 43 — Chemical Reactions
1) They are atoms joined together.
2) large
3) 54 321 (no atoms disappear, they are just rearranged)

Answers

Page 44 — Fuels
1) carbon and hydrogen
2) carbon dioxide and water
3) no (adding oxygen is an **oxidation** reaction)

Page 45 — Air Pollution — Carbon
1) carbon monoxide, CO
2) carbon particles / particulate carbon
3) It dissolves in rainwater and seas. Plants remove it from the air when they photosynthesise.

Page 46 — Air Pollution — Sulfur and Nitrogen
1) the air
2) acid rain

Page 47 — Reducing Pollution
1) They remove carbon monoxide and nitrogen monoxide from the gases the car gives out (exhaust gases).
2) They will fail their MOT test.
3) It means less fossil fuels are burned. So less carbon dioxide is given out.

Module C2 — Material Choices

Page 48 — Natural and Synthetic Materials
1) chemicals
2) Any three from: paper / cotton / wool / silk

Page 49 — Materials and Properties
1) tension and compressive strength
2) how difficult it is to cut
3) how much stuff there is in a certain amount of space

Page 50 — Materials, Properties and Uses
1) how good it is at the job it's supposed to do
2) Rubber. You want your tyres to be soft and bendy not hard and stiff.

Page 51 — Crude Oil
1) hydrogen and carbon
2) They are held together by forces.
3) Long hydrocarbons have lots of forces to keep the molecules together. So it takes lots of energy to break them apart.

Page 52 — Uses of Crude Oil
1) refining
2) true
3) Any two from: fuels / lubricants / raw materials

Page 53 — Polymerisation
1) small monomers
2) No. Different polymers can have very different properties.
3) Any one from: nylon / PVC

Page 54 — Structure and Properties of Polymers
1) a high melting point
2) yes
3) cross-linking / adding chemicals to hold the chains together

Page 55 — Nanotechnology
1) true
2) salt nanoparticles
3) The silver nanoparticles on the mask help to kill bacteria.

Module C3 — Chemicals in Our Lives

Page 56 — Tectonic Plates
1) tectonic plates
2) to work out how tectonic plates have moved
3) underwater

Page 57 — Resources in the Earth's Crust
1) limestone, salt and coal
2) sediment
3) salt

Page 58 — Salt
1) solution mining
2) Land can collapse into holes. Mining needs energy which comes from burning fossil fuels. This uses up resources and causes pollution.
3) by evaporating seawater

Page 59 — Salt in the Food Industry
1) It makes it tastier. It makes it last longer.
2) Any two from: it can give you high blood pressure. / It can increase the chance of getting stomach cancer. / It can increase the chance of getting osteoporosis. / It can increase the chance of getting kidney failure.
3) Any one from: the Department of Health / the Department for Environment, Food and Rural Affairs

Page 60 — Electrolysis of Salt Solution
1) brine / salty water
2) chlorine, hydrogen and sodium hydroxide
3) Electrolysis needs a lot of energy which comes from burning fossil fuels. Burning fossil fuels can create pollution.

Page 61 — Chlorination
1) Any two from: it stops algae growing. / It gets rid of bad tastes and smells. / It kills bacteria.
2) Chemicals in water can react with chlorine to make harmful chemicals.

Page 62 — Alkalis
1) salt and water
2) Any two from: to help dyes stick to cloth / to make glass / to make acidic soil neutral / to help turn fats and oil into soap
3) As industry became more important we needed more and more alkalis and so there weren't enough to go around.

Answers

Page 63 — Impacts of Chemical Production
1) because there are so many of them
2) carbon, hydrogen and chlorine
3) For example: animals could drink the water with the chemical in. If we then ate the animal and we would be taking in the chemical as well. / Chemicals eaten by animals could be passed along the food chain to humans.

Page 64 — Life Cycle Assessments
1) making the material, making the product, using the product, getting rid of the product
2) the use of resources, the energy used or made, how the environment is affected

Module P1 — The Earth in the Universe

Page 65 — The Solar System
1) a) the Sun
   b) planets
2) hydrogen atoms

Page 66 — Beyond the Solar System
1) the Milky Way
2) the diameter of the Milky Way

Page 67 — Looking Into Space
1) light and other electromagnetic radiation
2) It makes it hard to see dim objects in space.
3) false — We see stars as they were in the past.

Page 68 — The Life of the Universe
1) moving apart
2) 14 thousand million years ago
3) It's tricky to measure the very large distances in space and it's tricky to look at how things move in space.

Page 69 — The Changing Earth
1) Erosion is when rocks are broken into bits over a long time.
2) 4 thousand million years old

Page 70 — Wegener's Theory of Continental Drift
1) true
2) Any one from: he didn't have much evidence that he was right. / No one could detect the movement of the continents. / People had simpler ideas and explanations. / He wasn't a geologist.

Page 71 — The Structure of the Earth
1) the crust
2) earthquakes, mountains and volcanoes

Page 72 — Seismic Waves
1) true
2) P-waves

Page 73 — Waves — The Basics
1) true
2) hertz (Hz)
3) distance = speed × time = 3000 m/s × 60 s = **180 000 m**

Page 74 — Waves — The Basics
1) transverse
2) Speed = Frequency × Wavelength
   Speed = 16 000 Hz × 0.2 m = **3200 m/s**

Module P2 — Radiation and Life

Page 75 — Electromagnetic Radiation
1) photons
2) radio waves

Page 76 — Electromagnetic Radiation and Energy
1) 300 000 kilometres per second
2) the Sun

Page 77 — Ionising Radiation
1) ultraviolet radiation, x-rays, gamma rays
2) It damages them. It causes cell death and cancer.
3) lead, concrete and bone

Page 78 — Microwaves
1) water molecules
2) false — There isn't much evidence that mobile phones are dangerous.

Page 79 — Electromagnetic Radiation and the Atmosphere
1) carbon dioxide, water vapour and methane
2) true

Page 80 — The Carbon Cycle
1) Photosynthesis takes carbon dioxide from the air.
2) Respiration and burning add carbon dioxide to the air.

Page 81 — Global Warming and Climate Change
1) It will get worse. There will be more droughts and hurricanes.
2) It will rise.

Page 82 — Electromagnetic Waves and Communication
1) no
2) yes
3) light and infrared

Page 83 — Analogue and Digital Signals
1) false — Analogue signals can have any number.
2) less
3) the amount of information used to store pictures or sound

Answers

Module P3 — Sustainable Energy

Page 84 — Electrical Energy
1) true
2) 1000 J
3) power = voltage × current = 9 V × 17 A = **153 W**

Page 85 — Electrical Energy
1) Energy transferred = power × time
2) a) Energy transferred = power × time = 2 kW × 11 h = **22 KWh**
   b) Cost = number of kWh × cost per kWh
      = 22 kWh × 20p = 440p (= **£4.40**)

Page 86 — Efficiency
1) Any one from: light / sound
2) Useful energy = total energy − wasted energy
      = 300 000 J − 30 000 J = 270 000 J
   Efficiency = useful energy ÷ total energy × 100
      = 270 000 J ÷ 300 000 J × 100 = **90 %**

Page 87 — Sankey Diagrams
1) how much of the energy that goes into an appliance is turned into useful energy and how much is wasted energy
2) false — Appliances that are efficient have thick useful energy arrows.

Page 88 — Saving Energy
1) fossil fuels
2) Any two from: insulate your home / fit double glazing / fit draught-proofing / switch off appliances / turn down the heating / use energy efficient appliances

Page 89 — Energy Sources and Power Stations
1) It can travel long distances through the National Grid and can be used in many different ways.
2) Any two from: they produce a lot of energy. / They're pretty cheap. / They don't rely on the weather. / We've already got fossil fuel power stations so we don't have to build new ones.

Page 90 — Nuclear Energy
1) Any one from: it gives out a lot more energy than fossil fuels. / Nuclear fuel is fairly cheap. / It doesn't produce much carbon dioxide.
2) Any two from: it produces dangerous radioactive waste which is hard to get rid of. / Nuclear power stations take a long time to start up. / The overall cost is high. / It's non-renewable (will eventually run out).
3) Contamination is worse because you're exposed for a long time, which can cause more damage.

Page 91 — Wind and Solar Energy
1) Any one from: no pollution or carbon dioxide when you use them (only a bit when they're made). / No serious damage to the landscape. / No fuel costs and low running costs. / Renewable.
2) Any one from: start up costs are high. / Solar cells use quite a lot of energy to make. / It's hard to connect them to the National Grid. / They only work when it's sunny.

Page 92 — Wave and Tidal Energy
1) Any one from: wave turbines spoil the view. / Wave turbines are a danger to boats. / Wave turbines depend on the weather because waves need wind. / Start up costs are high.
2) Any one from: no pollution or carbon dioxide produced. / Tides don't depend on the weather. / Tides produce lots of energy. / There are no fuel costs and low running costs. / It is renewable.

Page 93 — Biofuels and Geothermal Energy
1) plants or rotting waste
2) Any one from: sometimes forests are chopped down and burnt to make space to grow biofuels. This means wildlife lose their habitats. / Rotting and burning forests gives off greenhouse gases.
3) Any one from: not much pollution / low running costs

Page 94 — Hydroelectricity and Reliable Fuel Supplies
1) Any one from: no pollution or carbon dioxide made when running. / Electricity can be made whenever it's needed. / It's fairly reliable. / No fuel costs and low running costs.
2) Any one from: high start up costs. / Rotting plants in the flooded valley give off carbon dioxide. / Wildlife can lose their habitats. / Reservoirs behind the dams can look ugly when they dry up.
3) to make sure the UK doesn't run out of electricity (no one source can supply enough power on its own)

Page 95 — Comparing Energy Resources
1) lowest
2) noise pollution

Page 96 — Generators and the National Grid
1) true
2) 230 V

Index

A

acid rain 46
adaptations (of organisms) 28
adult stem cells 17
air 41, 42
air pollution 45, 46
alkalis 62
alleles 10
amplitude 73
analogue signals 83
antibiotic resistance 21
antibodies 19, 20
antigens 19, 20
antimicrobials 21
arteries 23
asexual reproduction 16
asteroids 65
atmosphere 41, 42
atoms 43

B

bacteria 16, 18, 21
bar charts 98
Big Bang theory 68
biodegradable materials 40
biodiversity 32, 39
biofuels 93
blood vessels 23
brine 60
bulbs (plant) 16
burning 36, 44

C

capillaries 23
carbon cycle 36, 80
carriers 14
case study 101
catalytic converters 47
chemical reactions 43
chlorination 61
chlorine 60, 61
chromosomes 9, 10, 13
circulatory system 23
classification (of organisms) 32
climate change 81
clones 16
coal 44, 57
combustion 36, 44
comets 65
competition 35
conclusions 100
condensing 41
contamination 90

continental drift 70
core (of the Earth) 71
correlation 5, 99
crude oil 51, 52
crust (of the Earth) 56, 71
cystic fibrosis 14

D

Darwin, Charles 31
data 4, 98, 99
decomposers 36
density 49
digital signals 83
dimples 9
disease 18
DNA 9
dominant characteristics 11, 14
drug trials 22
drugs 15, 22

E

Earth 56, 71
earthquakes 72
ecstasy 27
efficiency (of appliances) 86
efficiency of energy transfer 34
eggs 10
electrical energy 84, 85
electrolysis 60
electromagnetic waves 75
embryonic stem cells 17
embryos 15, 17
employers 15
energy transfer (in ecosystems) 33, 34
environment 9, 16
environmental change 38
equations 8
erosion 69
evaluations 100
evolution 30, 31
extinction 30, 32, 35
eye colour 9

F

fair tests 5
false negatives 15
false positives 15
family trees 13
fibres 50
folding (of rock) 69
food chains 33
food webs 35

fossil fuels 89
fossil record 31
fossils 31, 56, 69
fractional distillation 52
fractions 52
frequency 73
fuels 44
functional proteins 9
fusion 65

G

galaxies 66
generators 96
genes 9, 10
genetic diagrams 11-13
genetic disorders 14
genetic testing 15
genetic variation 28
geothermal energy 93
global warming 81
greenfly 16
greenhouse effect 79

H

hazards 97
heart 23
heart attacks 24
heart disease 24
heart rate 24
Huntington's disease 14
hydrocarbons 51, 52
hydroelectricity 94
hydrogen 60
hypothesis 2

I

identical twins 16
immune system 19
inheritance 11
insurance companies 15
ionising radiation 77
irradiation 90

K

kidneys 26, 27
kilowatt-hours 85

Index

L

Lamarck 31
lichens 38
light pollution 67
light years 66
limestone 57
line graphs 99
living indicators 38
longitudinal waves 74

M

mantle 71
mayfly nymphs 38
mean (average) 98
melting point 49
memory cells 19, 20
method 97
microorganisms 18-20
microwaves 78
mining 58
miscarriage 15
molecules 43
monomers 53
mutations 28

N

nanoparticles 55
National Grid 96
natural selection 29
neutralisation 62
nitrates 37
nitrogen cycle 37
nitrogen oxides 46
non-living indicators 38
non-renewable energy sources 89
nuclear energy 90
nucleus (of a cell) 9

O

oceans 41
optical fibres 83
outliers 4
oxidation 44
ozone layer 79

P

parallax 67
parents 10
photosynthesis 33, 36
phytoplankton 38
placebos 22
plastic 50
plasticisers 63
pollutants 42
pollution 38, 45, 46
polymerisation 53
polymers 53, 54
power 84
power stations 89
prediction 2
proteins 9
pulse rate 24
Punnett squares 11, 12
P-waves 72

R

radiation 76
raw materials 48
recessive characteristics 11, 14
reduction 44
refining 52
reliable results 4
renewable energy sources 89
renewable materials 40
respiration 36
risk 6
rubber 50
runners (of plants) 16

S

salt 57-60
Sankey diagrams 87
scars 9
sediment 57
sedimentation 69
seismic waves 72
selective breeding 29
sex cells 10
sex chromosomes 13
sexual reproduction 10
sodium hydroxide 60
solar cells 91
solar system 65
specialised cells 17
species 28, 32
sperm 10
stem cells 17
stiffness 49
strength 49
structural proteins 9
sufferers 14
sulfur dioxide 46
sustainability 39, 40
S-waves 72
Sun 33, 65
synthetic materials 48

T

tectonic plates 56, 71
tidal power 92
transverse waves 74
twins 16

U

ultraviolet radiation 77
Universe 68
unspecialised cells 17

V

vaccination 20
variation 10, 28
veins 23
viruses 18, 21
volcanoes 41

W

water content (in the body) 26, 27
wave power 92
wave speed 73, 74
wavelength 73
Wegener 70
weight 9
wind power 91

X

X chromosomes 13
X-rays 77

Y

Y chromosomes 13